AF840189

BIOCHEMISTRY

M. Bilal Hanif

Toronto Academic Press

Biochemistry

M. Bilal Hanif

Toronto Academic Press
224 Shoreacres Road
Burlington, ON L7L 2H2
Canada
www.tap-books.com
Email: orders@arclereducation.com

ISBN: 978-1-77469-773-3

ABOUT THE AUTHOR

Muhammad Bilal Hanif is working as a research assistant at the Department of Inorganic Chemistry, Faculty of Natural Sciences, Comenius University in Bratislava, Slovakia. He obtained his graduation degree under a Chinese government scholarship at Xi'an Jiaotong University (XJTU), China. After obtaining his degree, He worked as a researcher at XJTU from 2018-2020. His research mainly focuses on solid oxide fuel cells, photocatalysis, and solid oxide electrolysis cells for energy conversion applications. He published articles in ACS Applied Materials & Interfaces; Chemical Engineering Journal; Energy & Environmental Materials and several high-quality journals articles. He has achieved Research Fund from Comenius University Bratislava, Slovakia for Young Scientists. He is also working as a Guest editor in Materials Journal.

Contents

3 Energy and Metabolism 47

4 Protein Synthesis and Folding 65

5 Gene Expression 87

List of Figures

List of Tables

PREFACE

Biochemistry is the branch of science that deals with the chemical processes and substances that occur within living organisms. It involves the study of the chemical properties and reactions of the biomolecules such as carbohydrates, lipids, proteins, and nucleic acids that make up living organisms. The principles of biochemistry underlie all life forms and provide a basis for understanding how living things function, grow, and evolve. The field of biochemistry is essential for the development of new drugs, therapies, and diagnostic tools, and it plays a crucial role in the advancement of biotechnology, genetics, and other life sciences. With its interdisciplinary approach and focus on the molecular basis of life, biochemistry is a fascinating and constantly evolving field that continues to reveal new insights into the workings of living organisms.

Chapter one provides an introduction to the basic principles of biochemistry, including the scientific method, the nature of life, and the organization of biological molecules. It covers topics such as the properties of biomolecules, cellular structures, and the organization of living systems. Chapter two focuses on biomolecules, examining the fundamental concepts of carbohydrates, lipids, proteins, and nucleic acids. It covers topics such as the structure and function of these biomolecules, and their roles in biological systems.

Chapter three delves into energy and metabolism, examining the basic principles of bioenergetics and the behavior of energy in biological reactions. It covers topics such as ATP synthesis, the laws of thermodynamics, and energy transfer in biological systems. Chapter four explores protein synthesis and folding, examining the structures, properties, and functions of proteins, from amino acid sequencing to protein folding and misfolding. It covers topics such as protein synthesis, post-translational modifications, and protein misfolding diseases.

Chapter five focuses on gene expression, examining the fundamental concepts of transcription, translation, and regulation of gene expression. It covers topics such as gene structure, DNA replication, and epigenetics. Chapter six covers DNA and RNA, examining the structures, properties, and functions of these nucleic acids. It covers topics such as DNA replication, transcription, translation, and RNA processing.

Chapter seven delves into metabolic pathways in the human body, examining the chemical reactions and processes that occur in living systems to produce and utilize energy. It covers topics such as the metabolism of carbohydrates, lipids, and amino acids, and the regulation of metabolic pathways. Chapter eight covers biochemical technologies, examining the applications of biochemistry in medicine, biotechnology, and other fields. It covers topics such as bioreactors, genetic engineering, and molecular diagnostics.

The book is written in an accessible and engaging style, with clear explanations and numerous examples and illustrations to help readers grasp complex concepts. The authors have drawn on their extensive experience in the field of biochemistry to provide a comprehensive and up-to-date resource for anyone interested in this fascinating subject. Overall, Introduction to Biochemistry is an essential resource for anyone seeking to gain a deeper understanding of the fundamental principles that govern the behavior of living systems. Whether you are a student of biochemistry, a professional in the field, or simply someone with a passion for science, this book will provide you with the knowledge and skills you need to succeed in the study of biochemistry.

-Author

CHAPTER 1

INTRODUCTION TO BIOCHEMISTRY

UNIT INTRODUCTION

Biochemistry is the study of processes & substances that exist in animals, plants, & microorganisms, as well as alterations that take place during their growth & life. Because it focuses on the chemistry of life, it applies techniques from analytical, physical & organic chemists and physiologists who research the molecular basis of essential functions. All of the chemical reactions that take place inside of an organism are referred to collectively as its "metabolism", such as the creation of complex molecules needed for life-sustaining processes or the breakdown of chemicals to provide energy. These chemical reactions are dependent on the activity of organic catalysts called enzymes, which are dependent on the genetic machinery of the cell to survive. Consequently, it should come as no surprise that biochemistry is used to examine chemical alterations in drug action, disease, & other aspects of medicine, as well as in genetics, food & agriculture (Packer & Landvik, 1989).

Physiological chemistry and biological chemistry are two considerably older words that are interchangeable with the term "biochemistry." The phrase "molecular biology" is frequently used to refer to those areas of biochemistry which deal with chemistry & process of very big molecules (such as nucleic acids & proteins). The name "biochemistry" has only been used to refer to this branch of study since around 1900. However, its earliest records are far older; they are a part of the earliest records of both physiology & chemistry (Hugli, 1986).

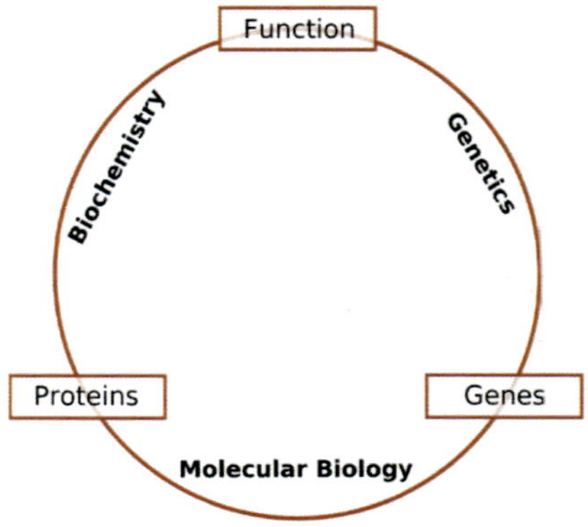

Figure 1.1. Illustration of biochemistry (Source:wikiwand, Creative Commons License).

Remember

Biochemistry is not just about memorizing the structures of molecules or the pathways of metabolism. It is the study of how life works at a molecular level and how these molecules interact with each other to carry out biological processes.

Learning Objectives

At the end of this chapter, the reader will be able to understand

- The historical background of biochemistry
- The areas of study for biochemistry
- The importance of nutrition
- The concept of digestion
- The significance of blood in the human body
- The methods of biochemistry

Key Terms

- Photosynthesis
- Biochemistry
- Adenosine triphosphate (ATP)
- Ribosomes
- Mitochondria
- Lysosomes
- Amino Acids

1.1. HISTORICAL BACKGROUND

Placing biological phenomena on solid chemical foundations has been a focus of the most major recent biochemistry studies.

However, chemistry has to overcome practical limitations and become a pure science in order to meaningfully impact agriculture and health. Beginning with Robert Boyle's work and ending with that of Antoine-Laurent Lavoisier, father of modern chemistry, this occurred between around 1650 and 1780. Boyle challenged tenets of the prevalent chemical theory of the time & professed that the true goal of chemistry was to ascertain the composition of things. John Mayow, one of his contemporaries, noticed the essential similarity between an animal's respiration & the process of burning, or oxidation, of organic materials in the atmosphere (Kohler, 1975). Lavoisier then quantitatively demonstrated the similarities between chemical oxidation and the respiratory process after conducting his foundational research on chemical oxidation and understanding the real nature of the process. In the late 18th century, chemists became interested in photosynthesis as another biological phenomenon. A turning point in the evolution of biochemical theory was shown by Joseph Priestley, Jan Ingenhousz, & Jean Senebier that photosynthesis is fundamentally the opposite of respiration.

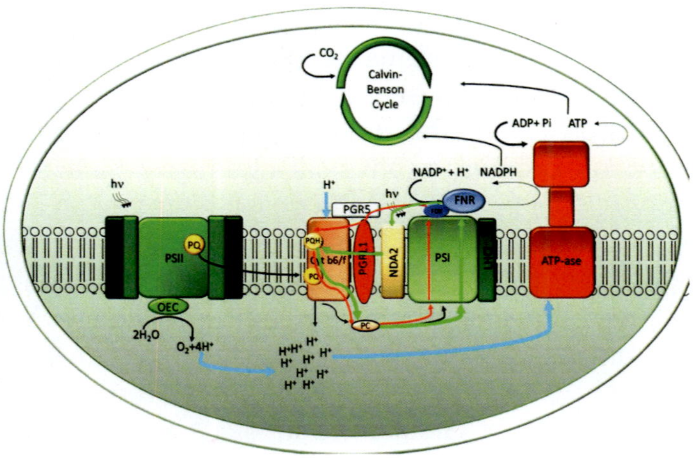

Figure 1.2. Schematic of the process of photosynthesis (Source: Vecchi et al., Creative Commons License).

Despite these early foundational findings, organic structure chemistry, one of the greatest achievements of 19th-century science, had to develop before biochemistry could advance quickly. There are thousands of different chemical compounds found in every living thing. A key challenge in biochemistry is understanding the molecular changes these chemicals undergo inside the live cell. The cellular mechanisms by which organic chemicals are synthesized and degraded had to be studied before determining the molecular structure of those substances present in live cells (Varki & Sharon, 2009).

Science has few well-defined divisions, and the divisions between organic and physical chemistry and biochemistry have always shown significant overlap. Physical and organic chemistry theories and methodologies have been appropriated by biochemistry and applied to physiological issues. The assumption is that matter underwent changes in living things and was not governed by the same chemical and physical laws applied to inanimate objects. As a result, these "vital" phenomena could not be explained in terms of ordinary chemistry or physics. The vitalists held this belief that natural products made by living creatures could never be synthesized using standard chemical processes (Fruton, 1976). Friedrich Wöhler's 1828 invention of the first laboratory synthesis of an organic substance, urea, was a setback for vitalists but not a big one. They turned to new lines of defense, claiming that urea was just an excretory chemical and not a byproduct of synthesis but a breakdown product. The organic chemists' success in synthesizing numerous natural compounds prompted the vitalists to retreat. Modern biochemistry takes it for granted that living cells follow the same chemical principles that apply to inanimate objects.

KEYWORD

Biochemistry explores chemical processes related to living organisms. It is a laboratory-based science combining biology and chemistry.

The practical requirements of man worked to advance the new science even as an unwarranted regard for living things was slowing down progress. The requirements of physicians, pharmacists, & agriculturalists provided an ever-present incentive for applying new chemical findings to several pressing practical issues during the 19th century, as organic & physical chemistry built an imposing body of theory (Kohler, 1973). A successful integration of chemistry with the study of biology was most prominently dramatized by two notable figures from the 19th century, Justus von Liebig & Louis Pasteur. When studying chemistry in Paris, Liebig brought his inspiration from interactions with Lavoisier's former students and coworkers back to Germany. He founded a world-class teaching and research facility in Giessen, one of its kind, which attracted students from around Europe.

Along with giving organic chemistry a solid foundation, Liebig published a lot of works, bringing it to the notice of other scientists and making it accessible to the general public. His classic works, which were published in the 1840s, had a significant impact on modern philosophy. Liebig described extensive chemical cycles in nature. He made the point that if it weren't for plants that can photosynthesize, animals would vanish off the face of the planet since they need the complex organic chemicals that only plants can produce for their nutrition (Harms & Thornton, 2013). Animal excretions & animal bodies after death are likewise broken down into basic substances exclusively useful to plants. Unlike animals, green plants need carbon dioxide, sunlight, water, & mineral salts to flourish. The soil must provide the minerals, & its fertility depends on its capacity to provide plants with these crucial nutrients. However, the removal of succeeding crops depletes the soil of these components, necessitating fertilizers. Liebig observed that a chemical study of plants might determine fertilizer composition. As a result, agricultural chemistry emerged as an applied science.

Liebig had less success in examining fermentation, putrefaction, and infectious illness. He acknowledged the similarities between these phenomena but was unwilling to accept that living things may be the cause. It was up to Pasteur to make that point clear. Pasteur discovered that different yeasts & bacteria were the cause of "ferments," or substances that induced fermentation &, in some cases, sickness, in the 1860s. He was the inventor of what is now known as bacteriology and established the value of using chemical techniques to investigate these small organisms (Rhee et al., 2022).

KEYWORD

Bacteriology is the branch and specialty of biology that studies the morphology, ecology, genetics and biochemistry of bacteria as well as many other aspects related to them.

Pasteur's ferments were later classified as enzymes in 1877. The German chemist E. Buchner demonstrated unequivocally in 1897 that fermentation could take place in pressed juice of yeast that contained no living cells. As a result, analysis reduced a live system of enzymes to a nonliving process of cells. Until 1926, when the first pure crystalline enzyme (urease) was discovered, the chemical makeup of enzymes was unknown. This enzyme and numerous others discovered to be proteins were known to be high-molecular-weight chains of smaller units termed amino acids.

When riboflavin (vitamin B2) was discovered to be a crucial component of an enzyme in 1935, the enigma of how trace amounts of food chemicals known as vitamins prevent illnesses like scurvy, beriberi, & pellagra became obvious. The idea that many vitamins are necessary for the cell's chemical reactions because of their function in enzymes has been supported by further research. Adenosine triphosphate (ATP), a compound, was first isolated from muscle in 1929. Work later showed that the cell's respiratory (oxidative) activities were connected to ATP synthesis. F.A. Lipmann first put forth the idea that ATP serves as the primary means of energy exchange in most cells in 1940.

It has been demonstrated that ATP is the main energy source for muscle contraction (Mendelsohn, 1964).

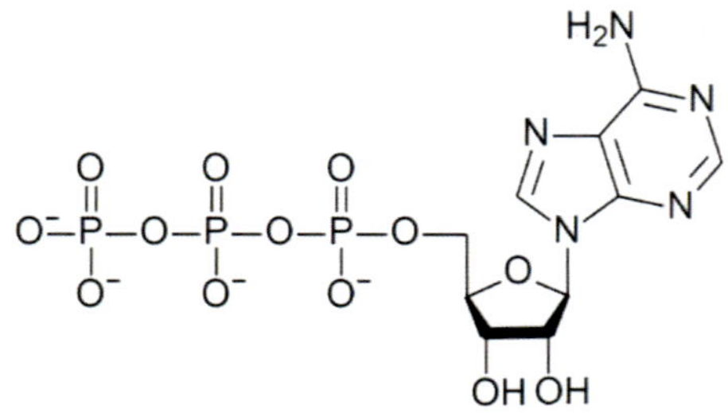

Figure 1.3. Illustration of Adenosine triphosphate (ATP) compound (Source: wikimedia, Creative Commons License).

Two American scientists, R. Schoenheimer and D. Rittenberg, first used radioactive isotopes of chemical elements to track the course of chemicals in the animal body in 1935. One of the most crucial instruments for examining the intricate chemical alterations during life processes was made possible by that method. Around the same time, other researchers used inventive technical innovations to pinpoint the locations of metabolic events through studies of cell mixtures, individual cells, tissue slices, organs, & specific cell components like nuclei, ribosomes, mitochondria, lysosomes, & membranes (Bing, 1971).

Deoxyribonucleic acid (DNA), first isolated from nuclei of pus cells in 1869 & given the name nucleic acid, was not recognized as a genetic material until 1944, when it was demonstrated that bacterial DNA could alter the genetic makeup of other bacterial cells. Within a decade of that finding, Watson & Crick postulated the double helix structure of DNA, giving scientists a solid foundation for understanding how DNA is involved in cell division & preserving hereditary traits. First chemical synthesis of protein, comprehensive mapping of molecular structures in some enzymes, & clarification of complex metabolic regulation systems, including molecular action of hormones, are only a few significant developments that have occurred since that period (Corbridge, 2013).

1.2. AREAS OF STUDY

All of the intricately linked chemical alterations inside of cells, also referred to as intermediate metabolism, are described when life is represented at the molecular level. Growth, reproduction, and heredity—all of which pique the interest of biochemists—are closely linked to intermediary metabolism & cannot be understood without them. A complex multicellular organism's traits and abilities may be broken down into the traits of its cells, & behavior of each cell can be described in terms of its chemical makeup & chemical changes taking on inside that cell (Sengupta, 1973).

Many organic compounds, or chemicals made of carbon coupled with different hydrogen & typically also oxygen atoms, are found in every living cell in addition to minerals, salts & water. Sulfur, phosphorus, and nitrogen are also typical components. In general, the majority of a cell's organic material can be divided into

- Protein,
- Carbohydrate, and
- Fat, Or Lipid.

Important components include nucleic acids and several other chemical compounds. There is a wide variety of unique chemicals in each class. Several compounds that occur but do not fit into the above-mentioned categories are typically in much smaller concentrations.

KEYWORD

Nucleic acids are biopolymers, macromolecules, essential to all known forms of life. They are composed of nucleotides, which are the monomer components: a 5-carbon sugar, a phosphate group and a nitrogenous base.

Proteins are important to life since they serve as key biocatalysts, structural components (like collagen), and defense mechanisms (like antibodies) against invading destructive forces. The work of German scientist Emil Fischer, whose findings from 1882 showed that proteins are very massive molecules, or polymers, constructed of around 24 amino acids, is the foundation for understanding the chemistry of proteins. Proteins can range in size from tiny molecules, like insulin, with a molecular weight of 5,700 (calculated using the weight of one hydrogen atom), to huge molecules, more than one million. In the 1950s, the insulin molecule's first exact amino acid sequence was discovered (Di Giulio et al., 1989). Using the potent physical methods of X-ray-diffraction analysis, it was also possible to detect amino acids chain in protein enzyme ribonuclease (molecular weight 12,700) by 1963. Nobel Prize winners J.C. Kendrew & M.F. Perutz created intricate molecular models of proteins hemoglobin & myoglobin (respiratory pigment in muscle) using X-ray investigations in the 1960s. Cutting-edge chemical studies later validated these models. Because chemical group arrangements in space provide crucial hints about the biological function of molecules, biochemists have a persistent interest in the structure of proteins.

The elements that make up carbohydrates include sugars, starches, and cellulose. Understanding how living cells deal with tiny molecules, such as carbohydrates, was remarkably advanced in the second quarter of the 20th century. During this time, the metabolism of carbohydrates was defined, and complex cycles of carbohydrate breakdown and subsequent storage and utilization were gradually described. (e.g., the Embden–Meyerhof glycolytic and Krebs cycles). By the 1950s, how carbohydrates affect respiration & muscle contraction was widely understood. The studies are still being improved (Wood, 1990).

KEYWORD

Metabolism is the process by which the body changes food and drink into energy. During this process, calories in food and drinks mix with oxygen to make the energy the body needs.

A diverse set of organic compounds known as fats, or lipids, can be extracted from biological material using nonpolar solvents, including ether, ethanol, & benzene. In the 1850s, groundbreaking research was completed on how carbs might cause the body to store fat. These investigations and later supporting data have demonstrated that the body continuously converts carbohydrates to fat. The primary location for fat metabolism is the liver. Even in the 1930s, research on fat absorption in the intestine was being done. It is well known that bile salts and pancreatic secretions work together to regulate fat absorption. Many biochemical studies focus on abnormalities in fat metabolism, which cause illnesses like obesity and other clinical problems. The link between high blood fat levels and arteriosclerosis ("hardening of the arteries") is also intriguing to biochemists (Rodwell et al., 2015).

The cells of all living things and viruses contain enormous, complex molecules with a very high molecular weight called nucleic acids. They play a crucial role in the production of proteins and in passing genetic information from one generation to the next. After their isolation in 1869, when they were first discovered as components of cell nuclei (hence their name), it was believed for many years that they were unique and could only be found there. This presumption was not substantially contested until the 1940s when it was discovered that there are 2 different types of nucleic acids: ribonucleic acid (RNA), which is found in all cells' cytoplasm and most viruses, and deoxyribonucleic acid (DNA), which is found in nuclei of all cells & certain viruses (Rose, 1981).

Throughout the 1940s & 1950s, the deep biological relevance of nucleic acids progressively became apparent. The process by which nucleic acids regulate protein synthesis and genetic transmission came to light.. The goal of research conducted in the 1960s was to improve the genetic code. In the late 1960s and early 1970s, promising efforts were undertaken to duplicate nucleic acid molecules outside cells or in a laboratory. By the middle of the 1980s, *in vitro* fertilization and DNA recombination (also known as "gene splicing") were two things that genetic engineering techniques had achieved (Koonin, 2009).

1.3. NUTRITION

Since the beginning of time, biochemists have been fascinated by the chemical components of animal diets. In addition to nutrients & water, all creatures must eat organic material. This organic material must be sufficient to meet animals' caloric or energy needs. Carbohydrates, protein & fat can all be utilized interchangeably for this purpose within limited bounds. Animals, however, also have dietary needs for particular chemical molecules. Many higher animals need specific necessary fatty acids, roughly ten different amino acids (so-called essential amino acids), & vitamins. But different animals have similar but not equal nutritional needs. Humans and guinea pigs need vitamin C or ascorbic acid, but rats do not (Friedman, 1999).

TIPS

Focus on consuming whole foods that are minimally processed and high in nutrients. This means eating plenty of fruits, vegetables, whole grains, lean proteins, and healthy fats. Avoid foods that are high in added sugars, saturated fats, and sodium.

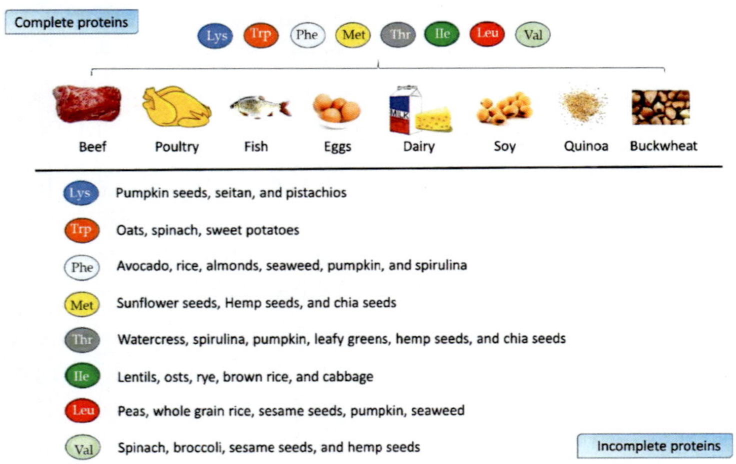

Figure 1.4. Sources of essential amino acids (Source: Lv et al., Creative Commons License).

Following the plant studies of the late 1700s, it was quickly realized that plants differ from animals because they do not require presynthesized organic material. Photosynthesis is the process by which green plants create all their cellular components from simple elements like water, salts, carbon dioxide & a nitrogen supply like nitrate or ammonia. As the name suggests, light is needed as an energy source, & sunlight typically provides it. Animals that consume plant carbohydrates can make fat from the process, the primary focus of which is the production of carbohydrates. Carbohydrates can also be converted into protein if ammonia is available. Despite what appears to be a significant disparity between plants' and animals' dietary needs, cells exhibit the same patterns of chemical change. The plant produces all the components it requires, but they are essentially the same as those used by animal cells and are frequently handled in the same manner after being generated. Plants could not meet animals' nutritional needs if their cellular components were not identical (Innis, 2014).

1.4. DIGESTION

KEYWORD

Proteolytic enzymes break down proteins in the body or on the skin. This might help with digestion or with the breakdown of proteins involved in swelling and pain. Some proteolytic enzymes that may be found in supplements include bromelain, chymotrypsin, ficin, papain, serrapeptase, and trypsin.

Large molecules comprise a portion of the organic food consumed by humans and other animals. These compounds are hydrolyzed, or broken down, to their constituent building blocks, in the digestive systems of higher animals. Polysaccharides are transformed into monosaccharides, and proteins are transformed into mixes of amino acids. All living things generally employ the same little molecules, but each species has its unique set of huge complex molecules. Because of this, an animal cannot utilize a plant's or another animal's protein directly; instead, it must first be broken down into amino acids. Then, amino acids are combined to form the animal's unique, distinctive proteins (Terra & Ferreira, 2012).

Additionally, food material must be hydrolyzed to become soluble compounds that may be absorbed. Long before the advent of modern chemistry, the liquefaction of stomach contents piqued the interest of early observers, and the hydrolytic enzymes secreted into the digestive tract were among the first to be thoroughly examined. The proteolytic enzymes found in gastric and pancreatic juice, respectively pepsin and trypsin, are still being thoroughly studied (Vashisht et al., 2017).

The byproducts of an the enzymatic action on its meal are absorbed through the intestinal wall & transported throughout the body through blood & lymph in an animal. Species without digestive tracts must also absorb environmental substances. In some cases, it appears that simple diffusion is enough to account for the transport of a material across a cell membrane. However, in some situations, such as when glucose is transferred from the intestine's lumen to the blood, the transfer occurs against a concentration gradient. In other words, glucose concentration may change from one location to another (Staggers et al., 1990).

It has been demonstrated that active secretion, or the metabolism of the tissue's respiratory system, is necessary to produce hydrochloric acid into gastric juice. The same is true for the process of salt absorption by the roots of plants. The energy produced during tissue oxidation must be captured to provide the energy required for secretion or absorption. A unique chemical coupling method is used to accomplish this. The biochemist's goal has been to clarify the nature of such coupling systems (Heimer et al., 1967).

1.5. BLOOD

One animal tissue that has always caused particular curiosity is blood. Since the beginning of biochemistry, blood has been the subject of much research, and its chemical makeup is better understood than any other tissue in the body. Since these exhibit distinctive changes in sickness, the doctor collects blood samples to determine blood sugar, urea, or inorganic-ion composition. Hemoglobin has been the subject of extensive research (Mann, 1999). The red blood corpuscles have hemoglobin, which transports oxygen from the lungs to the tissues. In the lungs, where there's high oxygen concentration, it mixes with oxygen and releases oxygen. In the tissues where there is a low oxygen concentration, it releases oxygen. Higher animals have hemoglobin that is related but not identical. Other pigments may replace and serve same purpose as hemoglobin in invertebrates. An intriguing section of biochemical research is the comparison of various substances.

Extensive research has also been done on blood plasma proteins. The blood's antibodies are found in gamma-globulin fraction of the plasma proteins, which has application as a vaccine. A disease-resistant animal produces antibodies well. Proteins called antibodies can bind to an antigen. (i.e., an agent that induces their formation). The antibody can shield an organism from infection by this substance when It is part of bacterium that causes disease. Immunochemistry is chemical study of antibodies, antigens, & their interactions (Broos et al., 2012).

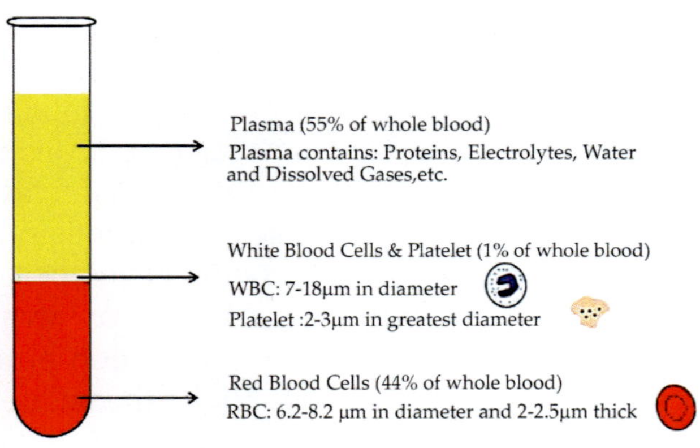

Figure 1.5. Schematic of components of blood (Source: Wang et al., Creative Commons License).

1.6. APPLIED BIOCHEMISTRY

Early biochemistry goals included developing analytical techniques for identifying different blood components since abnormal levels were thought to be a sign of metabolic disorders. The clinical chemistry laboratory is now an essential part of every hospital and serves as the doctor's primary investigative tool for diagnosing and treating disease. The most popular analytical techniques for diagnosing common diseases are still older, such as tests to measure levels of blood glucose in diabetes, levels of urea in kidney disease, levels of uric acid in gout, & levels of bilirubin in liver & gallbladder problems (Chibata & Wingard, 2014). With the development of enzyme studies, determination of certain enzymes, such as alkaline phosphatase, which is used to diagnose bone and liver disease, acid phosphatase, which is used to diagnose prostatic cancer, amylase, which is used to diagnose pancreatitis, and lactate dehydrogenase and transaminase, which are used to diagnose cardiac infarction, have diagnostic value. Plasma protein electrophoresis is frequently used to detect many cancers and liver illnesses. Diagnosing and evaluating treatment for atherosclerosis and heart disease increasingly use electrophoresis and ultracentrifugation of serum components (lipoproteins). Numerous specialized and complex techniques have been developed to meet the growing need for medical care, and devices have been made for the simultaneous automated examination of numerous distinct blood elements.

Did you get it?

The CRISPR-Cas9 gene editing technology has revolutionized biochemistry research and has been recognized with the Nobel Prize in Chemistry in 2020.

The food industry has also used analytical biochemical techniques to create crops with greater nutritional content & an ability to retain nutrients throughout food processing & preservation. Research in this field is focused on safeguarding vitamins, flavor, & color, which may all be lost if oxidative enzymes are present in preserved food. Enzyme tests are used to keep track of different phases of food preparation (Berlin, 1982).

The creation of novel medications has relied heavily on biochemical methods. Investigations on experimental animals and humans are used to test potentially beneficial medications to monitor the desired effects and look for harmful manifestations; these investigations heavily rely on the clinical biochemistry methods. A growing number of treatments have been developed as enzyme inhibitors to disrupt the metabolism of host or invasive agent. Still, many regularly used medications have been produced fairly empirically (through trial and error). Biochemical developments in our understanding of how natural hormones & antibiotics work promise to support the continued development of specialized medications (Hutchens, 1963).

1.7. METHODS IN BIOCHEMISTRY

Like other sciences, biochemistry strives to quantify or measure outcomes, occasionally using highly technical instruments. Analysis of the substances entering (foods, oxygen) and leaving an organism (excretion products, carbon dioxide) was the first method to analyze the occurrences in living things. This still serves as the foundation for so-called balancing tests on animals, in which, for instance, both food and excrement are carefully examined. Numerous chemical processes involving particular color reactions have been devised for this purpose, necessitating spectrophotometers for quantitative measurement. Oxygen and carbon dioxide measurements are frequently made using gasometric methods, resulting in respiratory quotients (ratio of carbon dioxide to oxygen) (Koch, 1953). Some further detail has been gathered by measuring the amounts of substances entering and exiting specific organs as well as by incubating slices of tissue in a physiological media outside of the body and examining the changes that occur in the medium. To better understand the chemistry of life, it was necessary to disrupt cellular structure (homogenization), isolate various enzymes & discrete chemical substances of a cell, and then separate various cell components, including nuclei, mitochondria, lysosomes, ribosomes, & membranes (Luo et al., 2004).

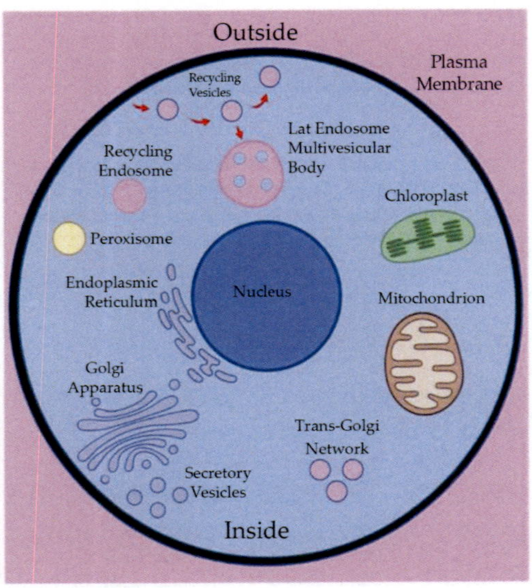

Figure 1.6. Image of cell components (Source: Baluška et al., Creative Commons License).

1.7.1. Centrifugation and Electrophoresis

The centrifuge is a crucial instrument in biochemical research because it rapidly spins suspended particles or even molecules in solution with high centrifugal forces, causing separations of such materials based on weight differences. Red blood cells and blood plasma can thus be separated, as can mitochondria and nuclei in cell homogenates and one protein from another in complex mixtures. Using suitable photography of the protein layers as they form in the centrifugal field, it is feasible to measure molecular weights of proteins. Proteins are separated by ultracentrifugation, which involves very fast spinning (Toyoda et al., 1978).

The electrical charge of biological molecules is another characteristic used for separation and analysis. Depending on how acidic the solution they are dissolved in, amino acids and proteins have net positive or negative charges. Such molecules adopt distinct migration rates toward positively (anode) or negatively (cathode) charged poles in an electric field, allowing separation. In liquids or when the proteins saturate a stationary media such as cellulose (filter paper), starch, or acrylamide gels, these divisions may take place. Counting the proteins in combination using the right color reactions of the proteins and scanning color intensities is possible. By using electrophoresis, distinct proteins can be extracted, identified, & purity of a particular protein can be assessed. (Electrophoresis of human hemoglobin, first conclusive demonstration of a "molecular illness, identified the aberrant hemoglobin in sickle-cell anemia") (Hadjiolov et al., 1966).

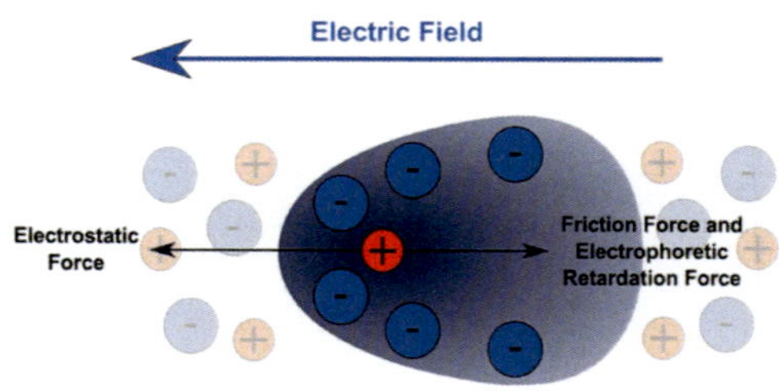

Figure 1.7. Illustration of electrophoresis (Source: Daniele Pugliesi, Creative Commons License).

1.7.2. Chromatography and Isotopes

Another foundation for analysis is the distinction between a substance's solubility in aqueous and organic solvents. In its former version, a separation was carried out in intricate apparatus by dividing material among several solvents. A streamlined version of the same concept gave rise to "paper chromatography," which allowed for the separation of small amounts of chemicals on filter paper and their identification by suitable color responses. Unlike electrophoresis, this approach has been used with various biological molecules and has significantly contributed to biochemistry research (Pitt, 2009).

The idea has been expanded from strips of filter paper to columns of various comparably inert media, enabling larger-scale separation & identification of closely connected biological molecules. The ability to determine proteins' precise amino acid composition has been made possible by separating amino acids by chromatography in columns of ion-exchange resins. After this assessment, different organic chemistry methods have been employed to clarify the precise order of amino acids in complex proteins. A different method of column chromatography relies on the relative rates at which different-sized molecules penetrate beads made of a complex carbohydrate. In contrast to smaller molecules, larger molecules are excluded & emerge first from a column of such beads. In addition to enabling the separation of biological components, this method also offers estimates of molecular weights. The use of isotopes (heavy or radioactive elements) in labeling biological substances & "tracing" their fate in metabolism may be the single most significant tool in understanding the complexity of metabolism. It has taken advanced mass spectroscopy and radioactive detection techniques to measure the isotope-labeled substances. Circular dichroism, electron spin spectroscopy, nuclear magnetic resonance, and X-ray crystallography are examples of other physical methods that have emerged as important tools for illuminating the connection between chemical structure and biological function (Cawood et al., 2005).

ACTIVITY 1.1.

In a biology lab, a student is tasked with identifying the characteristics of an unknown organism. The student knows that the organism is a multicellular eukaryote, but is not sure whether it is a plant or an animal. What fundamental characteristics of biology should the student use to differentiate between the two kingdoms and accurately classify the organism?.

SUMMARY

This chapter "Introduction to Biochemistry" provides an overview of the historical background and areas of study for biochemistry, as well as the importance of nutrition, digestion, blood in the human body, and methods used in biochemistry. The historical background of biochemistry includes the contributions of early scientists such as Antoine Lavoisier and Friedrich Wohler, as well as the development of modern biochemistry in the 20th century.

Biochemistry is a broad field that encompasses many areas of study, including the structure and function of biomolecules, metabolic pathways, and the regulation of gene expression. The importance of nutrition is highlighted in this chapter, including the role of essential nutrients and the impact of nutrient deficiencies on health. The concept of digestion is also discussed, including the various enzymes and organs involved in the breakdown and absorption of food.

The significance of blood in the human body is explored, including its role in transporting oxygen, nutrients, and waste products. Finally, the methods of biochemistry are outlined, including techniques such as chromatography, spectroscopy, and recombinant DNA technology. Overall, this chapter provides a solid foundation for understanding the fundamental concepts of biochemistry.

MULTIPLE CHOICE QUESTIONS

1. **Which of the following statements is true about the historical background of biochemistry?**
   a. Biochemistry originated in the 20th century
   b. Alchemists were the first to study biochemistry
   c. Biochemistry has its roots in the study of physiology and chemistry
   d. Biochemistry was primarily developed by philosophers
2. **Which of the following is NOT an area of study for biochemistry?**
   a. Metabolism
   b. Genetics
   c. Enzymology
   d. Immunology
3. **Why is nutrition important in biochemistry?**
   a. It is essential for the maintenance of human health
   b. It is a key factor in the prevention of diseases
   c. It provides the body with the necessary nutrients to carry out biochemical reactions
   d. All of the above

4. What is the concept of digestion?

a. The breakdown of large molecules into smaller ones for absorption by the body
b. The synthesis of large molecules from smaller ones for use by the body
c. The process of excreting waste products from the body
d. The conversion of nutrients into energy

5. What is the significance of blood in the human body?

a. It transports nutrients and oxygen to cells
b. It carries waste products away from cells
c. It helps regulate body temperature
d. All of the above

6. What are the methods used in biochemistry?

a. Spectroscopy
b. Chromatography
c. Electrophoresis
d. All of the above

Answers to Multiple Choice Questions

1. (c) 2. (b) 3. (d) 4. (a) 5. (d)

6. (d)

REVIEW QUESTIONS

1. What is the historical background of biochemistry? How did the field of biochemistry evolve over time?
2. What are the areas of study in biochemistry? Discuss the major sub-disciplines and their significance in understanding biological systems.
3. Why is nutrition important in biochemistry? How do nutrients provide the energy and building blocks required for biological processes?
4. What is the concept of digestion? Discuss the role of enzymes and other factors in breaking down food into its constituent molecules.
5. What is the significance of blood in the human body? How does blood carry out essential functions such as oxygen transport, immune defense, and waste removal?
6. What are the methods used in biochemistry research? Discuss the various techniques and instruments used to study biological molecules and processes.

REFERENCES

7. Berlin, N. I. (1982). Applied Biochemistry of Clinical Disorders. JAMA, 247(1), 88-88.
8. Bing, F. C. (1971). The history of the word 'metabolism'. Journal of the History of Medicine and Allied Sciences, 26(2), 158-180.
9. Broos, K., De Meyer, S. F., Feys, H. B., Vanhoorelbeke, K., & Deckmyn, H. (2012). Blood platelet biochemistry. Thrombosis Research, 129(3), 245-249.
10. Cantarow, A., & Trumper, M. (1963). Clinical biochemistry. Academic Medicine, 38(3), xviii-xlii.
11. Cawood, M. L., Field, H. P., Ford, C. G., Gillingwater, S., Kicman, A., Cowan, D., & Barth, J. H. (2005). Testosterone measurement by isotope-dilution liquid chromatography–tandem mass spectrometry: validation of a method for routine clinical practice. Clinical chemistry, 51(8), 1472-1479.
12. Chibata, I., & Wingard, L. B. (Eds.). (2014). Immobilized Microbial Cells: Applied Biochemistry and Bioengineering, Vol. 4. Elsevier.
13. Corbridge, D. E. (2013). Phosphorus: chemistry, biochemistry and technology, CRC PRESS, 1(1), 1-20.
14. Di Giulio, R. T., Washburn, P. C., Wenning, R. J., Winston, G. W., & Jewell, C. S. (1989). Biochemical responses in aquatic animals: a review of determinants of oxidative stress. Environmental Toxicology and Chemistry: An International Journal, 8(12), 1103-1123.
15. Ferry, J. G. (1992). Biochemistry of methanogenesis. Critical reviews in biochemistry and molecular biology, 27(6), 473-503.
16. Friedman, M. (1999). Chemistry, biochemistry, nutrition, and microbiology of lysinoalanine, lanthionine, and histidinoalanine in food and other proteins. Journal of Agricultural and Food Chemistry, 47(4), 1295-1319.
17. Fruton, J. S. (1976). The emergence of biochemistry. Science, 192(4237), 327-334.
18. Gold, H., Mirzaian, M., Dekker, N., Joao Ferraz, M., Lugtenburg, J., Codee, J. D., ... & Poorthuis, B. J. (2013). Quantification of globotriaosylsphingosine in plasma and urine of fabry patients by stable isotope ultraperformance liquid chromatography–tandem mass spectrometry. Clinical chemistry, 59(3), 547-556.
19. Granado-Lorencio, F., & Hernández-Alvarez, E. (2016). Functional foods and health effects: a nutritional biochemistry perspective. Current Medicinal Chemistry, 23(26), 2929-2957.
20. Hadjiolov, A. A., Venkov, P. V., & Tsanev, R. G. (1966). Ribonucleic acids fractionation by density-gradient centrifugation and by agar gel electrophoresis: a comparison. Analytical biochemistry, 17(2), 263-267.
21. Handelman, G. J. (2001). The evolving role of carotenoids in human biochemistry. Nutrition, 17(10), 818-822.
22. Harms, M. J., & Thornton, J. W. (2013). Evolutionary biochemistry: revealing the

historical and physical causes of protein properties. Nature Reviews Genetics, 14(8), 559-571.

23. Heimer, R., Schnoll, S. S., & Primack, A. (1967). Products of the peptic digestion of human γG-immunoglobulin. Biochemistry, 6(1), 127-134.
24. Hugli, T. E. (1986). Biochemistry and biology of anaphylatoxins. Complement, 3, 111-127.
25. Hutchens, T. T. (1963). Practical clinical biochemistry. JAMA, 184(11), 906-906.
26. Innis, S. M. (2014). Omega-3 fatty acid biochemistry: perspectives from human nutrition. Military medicine, 179(suppl_11), 82-87.
27. Koch, F. C. (1953). Practical methods in biochemistry. Practical methods in biochemistry., (6th Edit),1(1),1-20.
28. Kohler Jr, R. E. (1973). The enzyme theory and the origin of biochemistry. Isis, 64(2), 181-196.
29. Kohler, R. E. (1975). The history of biochemistry: a survey. Journal of the History of Biology,1(1), 275-318.
30. Koonin, E. V. (2009). Darwinian evolution in the light of genomics. Nucleic acids research, 37(4), 1011-1034.
31. Luo, G., Guenthner, T., Gan, L. S., & Humphreys, W. G. (2004). CYP3A4 induction by xenobiotics: biochemistry, experimental methods and impact on drug discovery and development. Current drug metabolism, 5(6), 483-505.
32. Mann, K. G. (1999). Biochemistry and physiology of blood coagulation. Thrombosis and haemostasis, 82(08), 165-174.
33. Marchetti, P. M., & Barth, J. H. (2013). Clinical biochemistry of dihydrotestosterone. Annals of Clinical Biochemistry, 50(2), 95-107.
34. Mendelsohn, E. (1964). The biological sciences in the nineteenth century: some problems and sources. History of Science, 3(1), 39-59.
35. Oi, T., Ogino, H., Hosoe, M., & Kakihana, H. (1992). Fractionation of strontium isotopes in cation-exchange chromatography. Separation Science and Technology, 27(5), 631-643.
36. Packer, L., & Landvik, S. (1989). Vitamin E: introduction to biochemistry and health benefits. Annals of the New York Academy of Sciences, 570, 1-6.
37. Pitt, J. J. (2009). Principles and applications of liquid chromatography-mass spectrometry in clinical biochemistry. The Clinical Biochemist Reviews, 30(1), 19.
38. Radman, R., Saez, T., Bucke, C., & Keshavarz, T. (2003). Elicitation of plants and microbial cell systems. Biotechnology and applied biochemistry, 37(1), 91-102.
39. Rhee, K. Y., Jansen, R. S., & Grundner, C. (2022). Activity-based annotation: the emergence of systems biochemistry. Trends in Biochemical Sciences, 1(1),1-20.
40. Rodwell, V. W., Bender, D. A., Botham, K. M., Kennelly, P. J., & Weil, A. P. (2015). Illustrated Biochemistry, 1 (1), (pp. 213-243).

41. Rose, S. P. R. (1981). What should a biochemistry of learning and memory be about?. Neuroscience, 6(5), 811-821.
42. Sengupta, I. N. (1973). Recent growth of the literature of biochemistry and changes in ranking of periodicals. Journal of Documentation, 29(2), 192-211.
43. Shapiro, L., & Weis, W. I. (2009). Structure and biochemistry of cadherins and catenins. Cold Spring Harbor perspectives in biology, 1(3), a003053.
44. Staggers, J. E., Hernell, O., Stafford, R. J., & Carey, M. C. (1990). Physical-chemical behavior of dietary and biliary lipids during intestinal digestion and absorption. 1. Phase behavior and aggregation states of model lipid systems patterned after aqueous duodenal contents of healthy adult human beings. Biochemistry, 29(8), 2028-2040.
45. Tanaka, N., Hosoya, K., Nomura, K., Yoshimura, T., Ohki, T., Yamaoka, R., ... & Araki, M. (1989). Separation of nitrogen and oxygen isotopes by liquid chromatography. Nature, 341(6244), 727-728.
46. Terra, W. R., & Ferreira, C. (2012). Biochemistry and molecular biology of digestion. In Insect molecular biology and biochemistry (pp. 365-418). Academic press. https://doi.org/10.1016/B978-0-12-384747-8.10011-X.
47. Thienpont, L. M., Van Uytfanghe, K., Blincko, S., Ramsay, C. S., Xie, H., Doss, R. C., ... & Sluss, P. M. (2008). State-of-the-art of serum testosterone measurement by isotope dilution–liquid chromatography–tandem mass spectrometry. Clinical chemistry, 54(8), 1290-1297.
48. Toyoda, H. I. R. O. O., Yamaguchi, K. A. Z. U. O., & Yoshikawa, H. I. R. O. S. H. I. (1978). Chromosome-membrane association in Bacillus subtilis. IV. Further purification of DNA-membrane complex by using a combination of centrifugation and electrophoresis. Journal of Bacteriology, 136(2), 799-802.
49. Varki, A., & Sharon, N. (2009). Historical background and overview. Essentials of Glycobiology. 2nd edition,1(1), page 1-20. Cold Spring Harbor Laboratory Press.
50. Vashisht, M., Rani, P., Onteru, S. K., & Singh, D. (2017). Curcumin encapsulated in milk exosomes resists human digestion and possesses enhanced intestinal permeability *in vitro*. Applied Biochemistry and Biotechnology, 183, 993-1007.
51. Wood, E. J. (1990). Biochemistry is a difficult subject for both student and teacher. Biochemical Education, 18(4), 170-172.

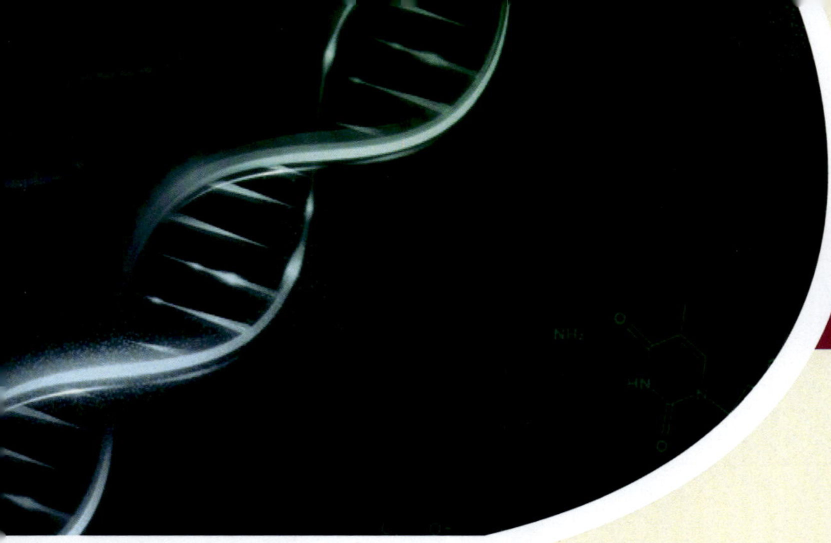

CHAPTER 2

BIOMOLECULES

UNIT INTRODUCTION

Biomolecules are molecules made by living things or substances that normally exist in animals and plants. Macromolecules or macromolecules like either primary or secondary metabolites could be present (Karplus & McCammon, 2002).

- Macromolecules
  - Nucleic acids
  - Lipids
  - Polysaccharides
  - Proteins
- Small Molecules
  - Terpenoids
  - Sugar Derivatives
  - Steroids
  - Alkaloids
- Amino Acids and Peptides (Bao, 2002).

TIPS

Understand that biomolecules such as proteins, carbohydrates, lipids, and nucleic acids are the building blocks of life. Take the time to learn how they are structured, how they function in the body, and how they interact with each other.

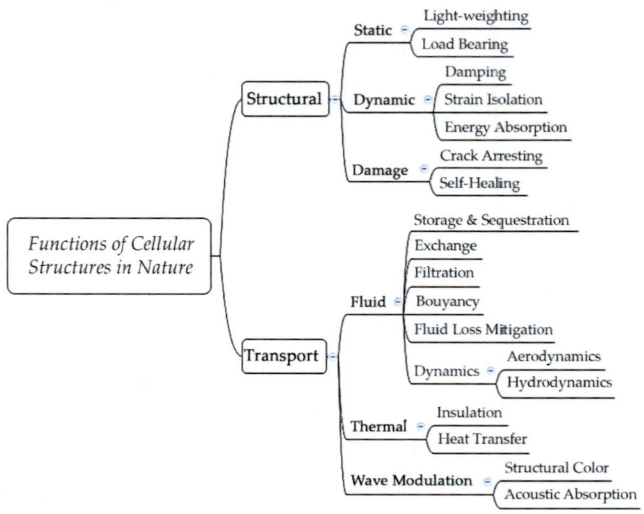

Figure 2.1. Cell structure functions (Source: Bhate et al., Creative Commons License).

Learning Objective

At the end of this chapter, the reader will be able to understand

- The importance of carbohydrates in the diet
- The significance of proteins
- The concept of lipids and their importance
- The roles of nucleic acids in the human body

Key Terms

- Carbohydrates
- Proteins
- Lipids
- Nucleic acids
- Biomolecules
- Monosaccharides
- Oligosaccharides
- Polysaccharides

2.1. CLASSIFICATION OF BIOMOLECULES

There are four fundamental groups of biomolecules:

- Carbohydrates
- Proteins
- Lipids
- Nucleic acids (Corriou et al., 1991).

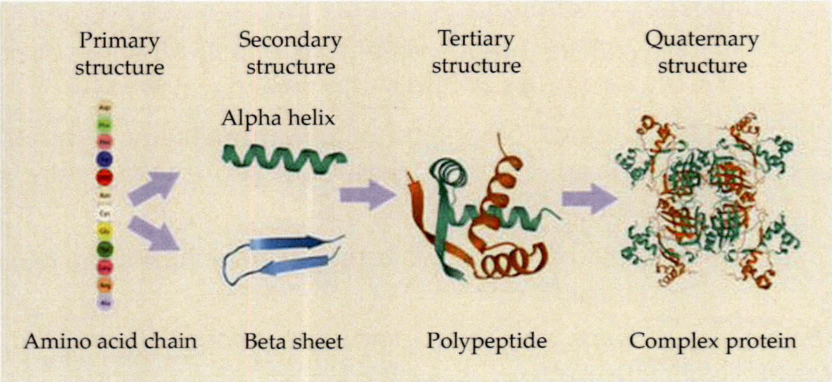

Figure 2.2. Illustration of classification of biomolecules with protein structures as an example (Source: Azulay et al., Creative Commons License).

2.1.1. Carbohydrates

A carbohydrate is a biomolecule consisting of carbon, hydrogen and oxygen atoms:

- Carbohydrates are a reliable energy source. Polysaccharides, or carbohydrates, are extended sugar chains. Simple sugars called monosaccharides have 3–7 carbon atoms in them.
- They are called reducing sugars because they have a free ketone or aldehyde group that functions as a reducing agent. Two monosaccharides combine to form disaccharides. The glycosidic bond links the two monosaccharides.
- Crystalline, Sweet, and water-soluble compounds, known as monosaccharides, are the building blocks of polysaccharides. They are complex carbohydrates that are not sweet. They do not have a crystalline structure and are water-insoluble.

Examples include glucose, starch, sucrose, cellulose, fructose, and maltose (Cummings & Stephen, 2007).

2.1.2. Lipids

- Lipids are long hydrocarbon chain molecules. Because

they can store a great deal of energy, lipids are energy storage molecules. Lipids are frequently fatty acid esters, the fundamental building blocks of biological membranes.

- Many lipid molecules have a tail (non-polar) and a head (polar). There are two forms of fatty acids: 1. Saturated, 2. Unsaturated fatty acids.
- Considering the kind of hydrophilic head that is present, the lipids found in biological membranes can be divided into three groups:
  - Glycolipids are a type of lipid with 1–15 sugar residues in the heads of their oligosaccharides.
  - The head of phospholipids (positively charged) is attached to the phosphate groups (negatively charged).
  - Sterols have a steroid ring in their head; an example is steroids.

Examples of lipids are *glycolipids*, *oils*, *phospholipids, fats,* etc (Sargent et al., 2003).

KEYWORD

Cytosine is one of the four nucleobases found in DNA and RNA, along with adenine, guanine, and thymine. It is a pyrimidine derivative, with a heterocyclic aromatic ring and two substituents attached.

2.1.3. Nucleic Acids

Nucleic acids are organic compounds with heterocyclic rings. The constituent parts of nucleic acids are called nucleotide polymers. Nucleotides comprise a phosphate group, a pentose sugar, and a nitrogenous base. A nucleoside is synthesized by joining a pentose sugar with a nitrogenous base. The nitrogenous bases are thiamine, adenine, guanine, cytosine, and uracil. The nucleotides that make up RNA and DNA have been polymerized (Fleck & Munro, 1966).

2.1.4. Proteins

Heteropolymers of amino acid chains make up proteins. The peptide bond, synthesized between the amino group and carboxyl group of succeeding amino acids, binds amino acids together. According to the number and sequence of the amino acid, proteins are formed of 20 distinct amino acids.

- Protein structure has four levels of configuration:
  - ***Primary structure*** -In this case, protein is a long amino acids chain ordered in a specific sequence. These are inactive proteins.

- ***Secondary structure***- Proteins comprise a long sequence of amino acids coiled and organized in a helix, wherein the amino acids connect by forming hydrogen bonds. The pleated sheet is the name of another structure, as in silk fibers (Helper & Gilman, 1992).
- ***Tertiary structure*** - The tertiary structure of proteins is synthesized when long polypeptide chains are folded and coiled and when ionic or hydrophobic connections or disulfide bridges are formed.
- ***Quaternary structure*** - The quaternary structure of a protein is defined as an assemblage of several polypeptides or independent subunits. Hemoglobin and insulin, as examples (Liu et al., 2016).

2.2. ROLES OF BIOMOLECULES

- The body uses **carbohydrates** as a food and energy source, and they also help our heart, brain, digestive system, neurological system, and immune system to function properly. Fatigue and poor mental function are caused by a diet low in carbohydrates.
- Each **protein** in the body serves a distinct purpose; for example, some proteins maintain the body's structure, provide mobility aid, and act as a barrier against pathogens and illnesses. Enzymes, antibodies, hormones, and contractile proteins are all examples of proteins.
- **Lipids** are primarily used by the body to store energy. Lipids make up structural membranes, a barrier that regulates material passage into and out of cells. Sterols and other lipid hormones aid in facilitating cell-to-cell communication.
- DNA and RNA are **nucleic acids** responsible for the cell's genetic makeup. Through the transcription and translation processes, they also assist in the synthesis of proteins (Aggarwal et al., 2000).

Did you Know?

The human body cannot produce enzymes that break down certain types of carbohydrates, such as fiber, leading to them being fermented by gut bacteria and producing important short-chain fatty acids.

2.3. CARBOHYDRATES

The only elements in carbohydrates are carbon, oxygen, and hydrogen, making them organic substances.

- The oxygen: hydrogen ratio is generally 2:1.

- Holds $\mathbf{C_n(H_2O)_n}$ as an empirical formula.
- These are hydrates of carbon; generally, they are known as polyhydroxy ketones and aldehydes.
- Carbohydrates are called saccharides, from a Greek word (sakkron), which means **sugar** (Kimura, 1996).

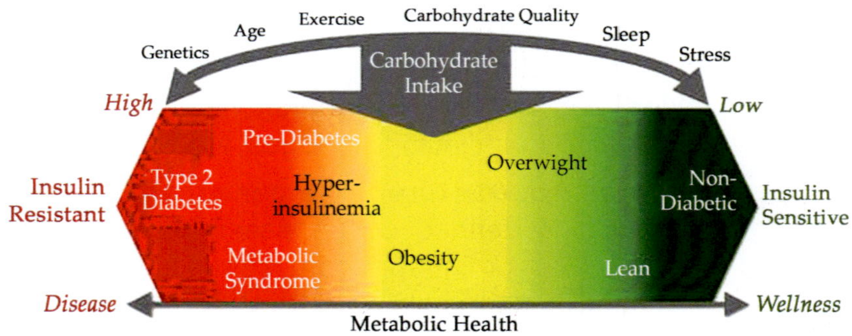

Figure 2.3. Schematic of the intake of carbohydrates (Source: Volek et al., Creative Commons License).

2.3.1. Nomenclature and Classification of Carbohydrates

Based on the hydrolytic products produced, carbohydrates are categorized into three main groups.

- *Monosaccharides*: The polyhydroxy aldehydes or ketones that make up the monosaccharides cannot be broken down by hydrolysis to produce simpler carbs. For example, Glucose, Galactose, Fructose, etc.
- *Oligosaccharides*: When the oligosaccharides are hydrolyzed, a specific number of monosaccharide molecules (2–9) are produced.
  - *Disaccharides* – On hydrolysis, they produce two monosaccharides molecules with molecular formula $C_{12}H_{22}O_{11}$; an example is maltose, sucrose, etc

$$\underset{\text{Sucrose}}{C_{12}H_{22}O_{11}} + H_2O \xrightarrow{H^+} \underset{\text{Fructose}}{C_6H_{12}O_6} + \underset{\text{Glucose}}{C_6H_{12}O_6} \,\vdots\, \underset{\text{Maltose}}{C_{12}H_{22}O_{11}} + H_2O \xrightarrow{H^+} \underset{\text{Glucose}}{2C_6H_{12}O_6}$$

- *Trisaccharides* – On hydrolysis, they produce three monosaccharide molecules with molecular formula $C_{18}H_{32}O_{16}$.

$$\underset{\text{Raffinose}}{C_{18}H_{32}O_{16}} + H_2O \xrightarrow{H^+} \underset{\text{Fructose}}{C_6H_{12}O_6} + \underset{\text{Glucose}}{C_6H_{12}O_6} + \underset{\text{Galactose}}{C_6H_{12}O_6}$$

- Tetrasaccharides – on hydrolysis, they produce four

monosaccharide molecules with molecular formula $C_{22}H_{42}O_{21}$. An example is Stachyose.

- *Polysaccharides: The carbohydrates with higher molecular mass produce several monosaccharides on hydrolysis. Examples are cellulose, glycogen, starch, dextrin, etc* (McNaught, 1997).

2.3.2. Comparison of Oligosaccharides, Monosaccharides, and Polysaccharides

Table 2.1. Tabular representation of a comparison of oligosaccharides, monosaccharides and polysaccharides (Source: Verbrugghe et al., Creative Commons License)

<table>
<tr><th colspan="5">CARBOHYDRATES</th></tr>
<tr><td colspan="2">SIMPLE SUGARS</td><td rowspan="4">OLIGO-SACCHARIDES
3-9 sugar units</td><td colspan="2">COMPLEX CARBOHYDRATES / POLYSACCHARIDES
>10 sugar units</td></tr>
<tr><td rowspan="3">MONO-SACCHARIDES / ABSORBABLE CARBOHYDRATES
1 sugar unit

Glucose
Fructose
Galactose</td><td rowspan="3">DI-SACCHARIDES
2 sugar units

Maltose
Lactose
Sucrose
Trehalose</td><td>Plant</td><td>Animal</td></tr>
<tr><td>Starch
α-glycosidic</td><td>Glycogen
α-glycosidic</td></tr>
<tr><td>Dietary fiber
β-glycosidic</td><td>Animal fiber
glycoproteins</td></tr>
<tr><td colspan="2">DIGESTIBLE CARBOHYDRATES</td><td colspan="3">INDIGESTIBLE CARBOHYDRATES</td></tr>
</table>

> **KEYWORD**
>
> **Genetic material** is the hereditary substance in the cell. It carries all information specific to an organism. It is known as DNA (deoxyribonucleic acid) or RNA (ribonucleic acid).

2.4. CHARACTERISTICS OF CARBOHYDRATES

- The structural framework of RNA and DNA, which makes up the genetic material, comprises the sugars ribose and deoxyribose.
- Bacterial and plant cell walls are made of polysaccharides, such as cellulose.
- Lipids and proteins connected to carbohydrates serve important functions in cellular interactions.
- Carbohydrates are hydroxylated ketones or aldehydes, which are organic molecules (Bergstrom & Helander, 2008).

2.4.1. Physical Characteristics of Carbohydrates

- Stereoisomerism: Compound shavings have a similar

structural formula, but their spatial arrangements vary.

- Example: Two isomers of glucose based on the penultimate carbon atom. They are D- and L-glucose, respectively.
- Optical feature: the circulation of plane-polarized light generating (+) glucose and (-) glucose.
- Diastereo isomers: the configurational difference concerning C2, C4, or C3 in glucose. Example: galactose, Mannose.
- Annomerism: the spatial arrangement concerning aldoses (first carbon) and ketoses (second carbon) (Li et al., 2018).

KEYWORD

Glucose is a sugar with the molecular formula $C_6H_{12}O_6$. Glucose is overall the most abundant monosaccharide, a subcategory of carbohydrates.

2.4.2. Chemical Characteristics of Carbohydrates

- Ozazone production with phenylhydrazine
- Reduction to alcohols
- Benedicts test
- Oxidation (Noike et al., 1985).

2.5. STRUCTURAL ORGANIZATION OF CARBOHYDRATES

The three structural configurations of carbohydrates are

- Open chain structure.
- Haworth structure.
- Hemi-acetal structure (Bucior & Burger, 2004).

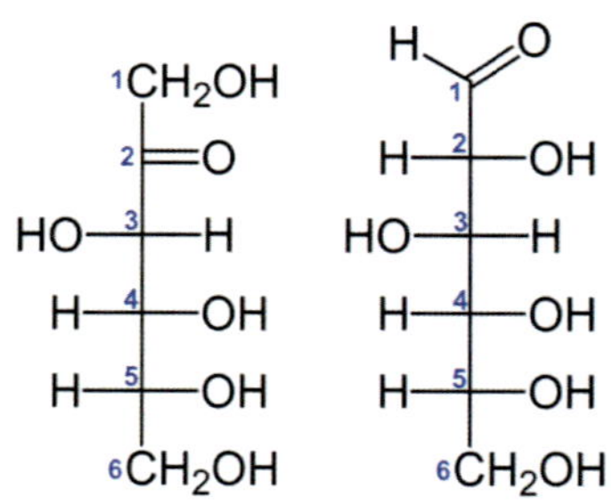

Figure 2.4. Structures of fructose (left) and glucose (right) (Source: wikimedia, Creative Commons License).

2.6. ROLES OF CARBOHYDRATES

- Carbohydrates are the main energy source in many animals because they immediately provide energy. The glycolysis/ Kreb cycle breaks down glucose to produce ATP.

- Glucose is the source of energy storage. Animals store it as glycogen, while plants store it as starch.
- The carbohydrates are intermediates during the production of lipids and proteins.
- Carbohydrates provide the brain with energy and help regulate nerve tissues.
- Antibiotics, surface antigens, vitamins, and carbohydrates bind to proteins and lipids to form receptor molecules (Weymouth, 1997).
- Similar to how plants and microbes generate structural and defensive constituents in their cell walls.
- They are important components of connective tissues in mammals.
- They participate in biological transportation, cellular communication, and growth factor stimulation.
- Fiber-rich carbohydrates aid in the prevention of constipation.
- Additionally, they support immune system activation (Thompson & Thorpe, 1987).

KEYWORD

Microbes are organisms that are too small to be seen without using a microscope, so they include things like bacteria, archaea, and single cell eukaryotes — cells that have a nucleus, like an amoeba or a paramecium.

2.7. CARBOHYDRATES WITH EXAMPLES

- ***Monosaccharides***: Galactose, ribulose, glycerose, glucose, ribose, erythrose, and fructose.
- ***Oligosaccharides***: Stachyose, Maltose, sucrose, lactose, and raffinose.
- ***Polysaccharides***: Glycogen, Starch, pectin, cellulose, hyaluronic acid, and inulin (Werz & Seeberger, 2005).

Table 2.2. Tabular representation of examples of carbohydrates (Source: Davidson, Creative Commons License)

Polysaccharide Names	**Occurrence**	**Composition**	**Functions**
Starch	In many plant species, as basic storage, carbohydrate	Glucose polymer has amylose with a straight chain of glucose and amylopectin with a branched chain of glucose.	Stores reserve food
Glycogen	Animals (equal to starch)	Glucose polymer	Stores of reserved food

Cellulose	Plant cell wall	Glucose polymer	Matrix of Cell Wall
Inulin	In tubers and roots (like Dahlia)	Fructose Polymer	Reserve food storage
Hemi cellulose	Plant cell wall	Pentoses and sugar acids polymer	Matrix of Cell wall
Pectin	Plant cell wall	Galactose and its derivatives polymer	Matrix of Cell wall
Chitin	The body wall of arthropods and fungi.	Glucose polymer	Exoskeleton Impermeable to water
Lignin	dead cells like sclerenchyma in Plant cell wall	Glucose polymer	Cell wall matrix
Hyaluronic acid	matrix of Connective tissue, Outer covering of mammalian eggs	Sugar acids polymer	Protection and Ground substance,
Murein	The cell wall of prokaryotic cells	Cross linkage of Polysaccharide with amino acids	Protection of structure
Gums and mucilages	Gums - bark or trees. Mucilages - flower	Sugars and sugar acids polymers	Retain water in dry seasons
Heparin	Connective tissue cells	Closely related to chondroitin	Anticoagulant

2.8. LIPIDS

KEYWORD

Hydrophobic means lacking an affinity for water; insoluble in water; repelling water.

Body lipids are normally migrated in plasma with protein affiliation, such as in the case of lipoprotein particles or on albumin, or organized, as in the case of lipids associated with membrane or triacylglycerol droplets in adipocytes. Lipids are a diverse category of chemical compounds that are hydrophobic and water-insoluble that can be drawn from tissues by nonpolar solvents.

- Lipids are the body's main energy source and the hydrophobic border.
- Lipids have extra roles in the body. For instance, some fat-soluble vitamins serve as coenzymes or regulators, and prostaglandins and steroid hormones are important regulators of homeostasis (Marsh & Weinstein, 1966).

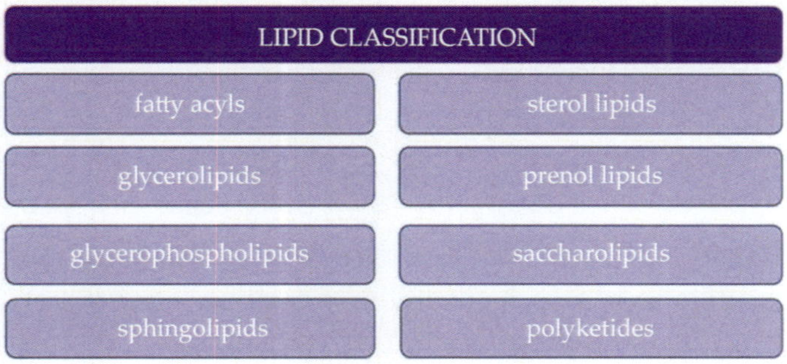

Figure 2.5. Schematic of the classification of lipids (Source: Nowowiejska et al., Creative Commons License).

2.8.1. General Properties of Lipids

- Water does not readily dissolve lipids.
- They are dispersible in non-polar solvents like methanol, ether, and chloroform.
- Lipids contain a lot of energy and are digested to produce calories.
- Lipids protect nerve axons and serve as insulators of electricity.
- Saturated fatty acids are present in fats and solids at room temperature. For instance, animal fats.
- Unsaturated plant fats are liquid at normal temperatures.
- Natural fats have no color and a very bland flavor (Janmey & Kinnunen, 2006).
- Due to their limited water solubility, lipids are called hydrophobic compounds.
- Organic solvents like ether, acetone, and benzene are easily soluble.
- The length of the fatty acid chain and the level of unsaturation both affect a substance's melting point.
- Geometric isomerism, also known as cis-trans isomerism, is caused by a double bond in the unsaturated fatty acid of the lipid molecule.
- Fats are poor heat conductors and have insulating properties.
- Emulsification is the procedure used to break up a large lipid bulk into numerous tiny droplets. Before the intestinal walls can consume the fats, they must be emulsified.

KEYWORD

Room temperature is a range of air temperatures that most people prefer for indoor settings. It feels comfortable to a person when they are wearing typical indoor clothing.

- The lipase enzyme breaks down the fats into fatty acids and glycerol.
- Saponification refers to the breakdown of lipids by an alkaline solution. As a consequence of this reaction, glycerol, and soap-like fatty acid salts are produced.
- The development of bacteria that release enzymes like lipases leads to hydrolytic rancidity. These separated the fats into free fatty acids and glycerol (Cronan & Gelmann, 1975).

Remember

While often thought of as simply a source of energy or a component of cell membranes, lipids are incredibly diverse and play many important roles in the body. For example, signaling lipids like prostaglandins and leukotrienes are involved in inflammation and immune responses, while sterols like cholesterol are precursors to many important hormones. Understanding the functions and roles of lipids in the body is important for developing treatments for a variety of diseases.

2.8.2. Lipid Classification

- **Simple lipids:** These are Esters of fatty acids with alcohols.
  - Esters of fatty acids linked with glycerol are **fats**. **Oils** are lipids in liquid form.
  - Esters of fatty acids linked with higher molecular monohydric alcohols are **waxes**.
- **Complex lipids:** Esters of fatty acids comprising groups, fatty acids, and alcohol.
  - *Phospholipids:* Lipids have a phosphoric acid residue, fatty acids, and alcohol (Fahy et al., 2011).
  - They usually contain nitrogen bases as well as other substituents. For instance, the alcohol in glycerophospholipids is glycerol, while the alcohol in sphingophospholipids is sphingosine.
  - *Glycolipids:* A fatty acid, carbohydrate, and sphingosine make up glycolipids.
  - *Complex lipids:* Amino lipids and sulfolipids are complex lipids. Lipoproteins might be part of this group.
- **Precursor and derived lipids***:* Fatty aldehydes, Steroids, glycerol, fatty acids, other alcohols, hydrocarbons, ketone bodies, hormones, and lipid-soluble vitamins are taken in this category (O'Keefe, 2002).

2.8.3. Essential Fatty Acids

- Two fatty acids are important in humans
  - **Linoleic acid**, the precursor of arachidonic acid, is the substrate for prostaglandin synthesis.

- α-linolenic acid is the precursor for development and growth.

- Scaly dermatitis and abnormalities of vision and neurons occurs due to the deficiency of essential fatty acids (Di Pasquale, 2009).

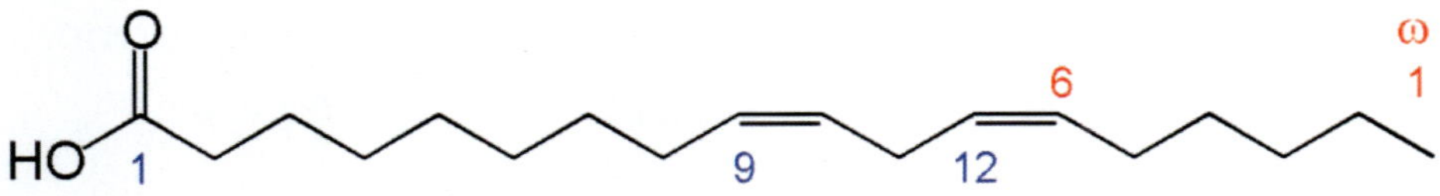

Figure 2.6. Illustration of linoleic acid (Source: wikmedia, Creative Commons License).

KEYWORD

High-density lipoprotein(HDL) cholesterol, sometimes called "good" cholesterol, absorbs cholesterol in the blood and carries it back to the liver. The liver then flushes it from the body. High levels of HDL cholesterol can lower your risk for heart disease and stroke.

2.8.4. Regulation of Blood Cholesterol Levels

- Because cholesterol and fats cannot be dissolved in the blood, they are packed with proteins (to create lipoproteins) for transportation.
- High-density lipoproteins (HDL),
- Low-density lipoproteins (LDL) transport cholesterol from the liver to the rest of the body, collect extra cholesterol, and transport it back to the liver for clearance (McGandy et al., 1972).
- Hence HDLs lower blood cholesterol levels, but LDLs improve blood cholesterol levels (the "bad" cholesterol).
- Consuming large amounts of particular fats will affect blood cholesterol levels differently.
  - Saturated fats raise LDL levels, which raise blood cholesterol levels.
  - Trans fats raise LDL and lower HDL levels, significantly raising blood cholesterol levels.
  - Unsaturated (cis) fats raise HDL levels, which lower blood cholesterol levels (Brown et al., 1981).

2.9. EXAMPLES OF LIPIDS

- ***Fatty acids*** - Oleic acid, Palmitoleic acid, Linoleic acid, Arachidonic acid.
- **Plant oils** - Corn, Peanut, Coconut oil, Palm, Sunflower oil.
- ***Phospholipids*** - Phosphatidyl inositols, Lecithins, Plasmoalogens, Cephalins, Sphingomyelins.

- ***Fats and Oils*** - Animal fats - Lard, Butter, Herring oil, Human fat.
- ***Terpenes*** - Monoterpenes, Diterpenes, Sesquiterpenes, Triterpenes.
- ***Waxes*** - Spermacti, Carnauba wax, Beeswax.
- ***Glycolipids*** - Nervon, Kerasin, Phrenosin, Oxynervon.
- ***Carotenoids***- Carotenes, Lycopene, Xanthophylls.
- ***Steroids*** – Cholesterol (Killian, 1988).

KEYWORD

Cholesterol is a waxy substance found in your blood. Your body needs cholesterol to build healthy cells, but high levels of cholesterol can increase your risk of heart disease.

2.10. BIOLOGICAL FUNCTIONS OF LIPIDS

- **Food material:** Lipids give rich calorific food. 1g lipid yields 9.3 kilocalories of heat.
- **Food reserve:** Because of their insolubility in aqueous media, lipids could be stored easily in the body as a food reserve.
- **Structural component:** Lipids are an important component of the cell membrane.
- **Heat insulation:** The fats are highlighted for their excellent insulation properties. Animals inhabiting cold regions and aquatic mammals like whales accumulate large amounts of fat in the subcutaneous layers.
- **Fatty acid absorption:** Phospholipids are involved in absorbing and transporting fatty acids (Muro et al., 2014).
- **Hormone synthesis:** Cholesterol, a steroidal lipid, is the starting material for synthesizing sex hormones, cholic acids, adrenocorticoids, and vitamin D.
- **Vitamin carriers:** Lipids behave as fat-soluble carriers like D, A, and E.
- **Lowering Blood cholesterol level:** Beef and chocolates, especially the latter, were long considered important causes of heart disease because of their high quantity of saturated fatty acids, which raise blood cholesterol levels and block arteries. In contrast, research by Scott Grundy and Andrea Bonanome (1988) at the University of Texas shows that a minimum of stearic acid, a saturated fatty acid vital to beef fat and cocoa butter, does not at all raise blood cholesterol levels. The researchers randomized 11 males to one of three cholesterol-deficient meals for three weeks each. Large amounts of palmitic acid, which tends to increase

cholesterol, were present in the original composition. There was a lot of oleic acid in the second recipe. Patients on the stearic acid diet had blood cholesterol levels that were 14% lower than those on a diet high in palmitic acid, and participants on the oleic acid diet had blood cholesterol levels that were 10% lower.

- **Antibiotic agent:** Squalamine, a shark blood steroid, has been discovered to possess potent antifungal and antibacterial effects. This seems to explain why infections and cancer are so rare in sharks (Wen et al., 2018).

KEYWORD

Immunoglobulins (Ig) or antibodies are glycoproteins produced by plasma cells. B cells are instructed by specific immunogens, for example, bacterial proteins, to differentiate into plasma cells.

2.11. PROTEINS

Proteins are substantial biomolecules, or macromolecules, of one or more extended amino acid chains. Proteins are regarded as the foundation of life. The most prevalent intracellular macromolecules are proteins. Skin, ligaments, hair, muscles, calluses, cartilage, and tendons give the body of multicellular organisms structure and protection. In various hormones, immunoglobulins, enzymes, etc., proteins govern and catalyze the body's chemical makeup. The building blocks of proteins are amino acids, synthesized by living things depending on the data in their DNA (Somero, 1995).

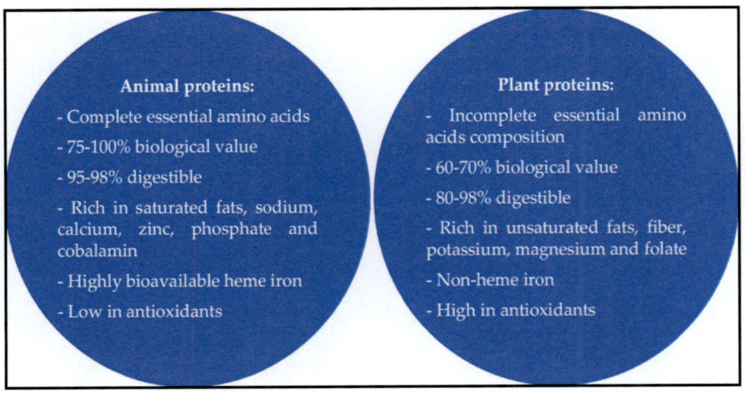

Figure 2.7. Image of essential protein (Source: Małecki et al., Creative Commons License).

General properties of Proteins

- The elements oxygen, nitrogen, carbon, and hydrogen are all components of proteins, which are organic compounds.
- The most significant biomolecules are proteins, which form the core of a cell's cytoplasm.
- The structural components of body tissues are proteins.

- Amino acids make up proteins.
- Proteins help the body develop and repair tissues and provide energy and heat (Anson & Mirsky, 1925).
- Because proteins can be consumed quickly, just small values are kept in the body.
- Proteins are sometimes called the "bricks" of the body since they form the muscles, bones, hair, and other tissues.
- Proteins, like enzymes, are active substances that participate in metabolic processes.
- Proteins are also used to make antibodies and blood hemoglobin. - The molecular weight of proteins ranges from 5 to 300 kilo-Daltons (Kinsella & Melachouris, 1976).

2.12. PROPERTIES OF PROTEINS

2.12.1. Physical properties

- Proteins are crystalline, tasteless, and colorless.
- Proteins have different shapes and might be simple (crystalloid) to long (fibrillar) structures.
- Proteins have two distinct structural patterns – 1. Globular proteins, 2. Fibrilar proteins.
- Plants possess globular proteins that are spherical. At the same time, animals possess Fibrilar proteins that are thread-like.
- Naturally, proteins have a larger molecular mass that ranges between 5×10^3 and 1×10^6.
- Because of their massive size, proteins show several colloidal characteristics.
- Proteins have extremely slow diffusion rates (Foegeding et al., 2006).
- They show the Tyndall effect.
- Proteins tend to denature, and the denaturation process is followed by coagulation.
- Denaturation might occur due to some chemical or physical agents. The physical agents include freezing, shaking, heating, etc. Chemical agents include radioactive, X-rays, and ultrasonic radiations.
- Proteins show amphoteric properties, i.e., they could be an Acid and Alkali.

- Due to an amphoteric nature, they could form salts with both anions and cations depending on the net charge.
- The pH affects the solubility of the proteins. The isoelectric point has the lowest solubility that rises with a rise in alkaline or acidic nature.
- Proteins exhibit the plane of polarized light to the left, **laevorotatory** (Wihodo & Moraru, 2013).

2.12.2. Chemical Properties

- Amino acids are produced as their hydrochlorides when proteins are digested by acidic substances such as strong HCl.
- On hydrolysis with alkaline agents, certain amino acids, such as cysteine, arginine, and serine, are hydrolyzed, and the amino acids' optical activity is destroyed.
- Esters are produced in a reaction of proteins with alcohols during the Esterification process.
- Amines and amino acids combine to generate amides.
- Acid salts are synthesized when free amino acids or proteins interact with acidic minerals like HCl.
- When an amino acid in an alkaline media combines with several acid chlorides, it results in an acylation reaction.
- Xanthoproteic test - When proteins are boiled with concentrated HNO_3, a benzene ring causes the color to turn yellow.
- The Folin's test, which is particular for the amino acid tyrosine, involves the development of blue color in an alkaline solution of phosphomolybdotungstic acid when a phenol group is present (Cozzone, 2002).

KEYWORD

Esterification is the process of combining an organic acid (RCOOH) with an alcohol (ROH) to form an ester (RCOOR) and water; or a chemical reaction resulting in the formation of at least one ester product.

2.13. STRUCTURAL ORGANIZATION OF PROTEINS

Only 20 distinct amino acids are polymerized into linear chains to produce proteins. L-amino acid polymers make up proteins. Proteins have a relatively intricate structure that is separated into four organization levels.

- **Primary structure:** The polypeptide chain is the linear arrangement of amino acids that makes up proteins.

Hexosaminidase and dystrophin are two examples of proteins that have a primary structure.

- **Secondary structure:** The spatial organization of proteins through polypeptide chain folding. Myoglobin is an example of a protein with a secondary structure.
- **Tertiary structure:** a protein's three-dimensional structure. The polypeptide chain is held in this final confirmation by a variety of forces:
  - Hydrogen Bonds
  - Disulfide Bonds
  - Van der Waals Forces
  - Polar/Nonpolar Interactions
  - Ionic Interactions
- **Quaternary structure:** Some proteins comprise subunits, groups of two or more polypeptide chains. Quaternary structure describes how these subunits are arranged spatially. DNA polymerase and ion channels are two proteins having quaternary structures (Crippen, 1978).

KEYWORD

DNA polymerase is a type of enzyme that is responsible for forming new copies of DNA, in the form of nucleic acid molecules.

An amide bond bonds amino acids to form proteins. Alkyl, alcoholic, aromatic, phenolic, thioether, or thiol, basic groups like amine or guanidine, and carboxylic acids are only possible R-groups. These distinct groups work together in proteins to carry out various tasks. The fundamental structure, a chain of amino acids, folds into secondary and tertiary structures through various interactions. Some proteins form quaternary structures when two or more of their tertiary protein subunits come together. The three-dimensional structure that a protein can assume significantly impacts how well it functions. Some proteins provide structural stability.

Glutathione – A functional tripeptide: Peptides are shorter chains of amino acids connected by amide bonds. A tripeptide called glutathione is synthesized from cysteine, glutamic acid, and glycine. The middle amino acid, cysteine with a free SH group, is essential for the process. Thiols are quickly converted to disulfides by oxidation, and glutathione will self-destruct if exposed to an oxidizing agent. Later, glutathione reduces back to its thiol form from its oxidized state.

Additionally, glutathione functions as a nucleophile, converting hazardous chemicals that enter the body into innocuous molecules (Yue & Dill, 1995).

2.14. ROLES OF PROTEINS

- Proteins comprise most of the body's non-skeletal framework, found in hair, muscles, skin, and other tissues. As an illustration, nails, and hair both contain the protein keratin.
- Some proteins act as hormones and control various bodily processes. As an illustration, consider the protein hormone insulin, which controls blood sugar levels.
- A few proteins function as enzymes, catalyzing or aiding biological processes. Pepsin and trypsin, as an example.
- Some proteins function as antibodies, defending the body against the effects of foreign organisms or substances.
- Proteins carry various chemicals in the blood of various tissues. A protein that transports oxygen is hemoglobin.
- Contractile proteins aid in muscle contraction and cells. A contractile protein is called myosin.
- The glycoprotein fibrinogen aids in the healing of wounds. It stops blood loss and hinders the spread of pathogens (Hamilton, 2007).

ACTIVITY 2.1.

An agricultural company wants to improve the nutritional value of its crops by enriching them with proteins. What biomolecules are essential for protein synthesis in plants, and how could the company use this knowledge to achieve their goal?

2.15. NUCLEIC ACIDS

The fact that nucleic acids (DNA and RNA) contain our genetic information makes them top-level molecules. They are polymers of nucleotide, and nucleotides are composed of a sugar, a heterocyclic base, and a phosphate ester.

Phosphates are important substances in nature as they help create stable connections between different molecules and can be multiplied to create reactive molecules. Adenosine triphosphate (ATP) is one of the most significant compounds in nature and the most significant of these nucleotides (Hutchison & Munro, 1961).

2.16. NUCLEIC ACIDS COMPONENTS

Only two sugars, five bases, and one phosphate group can be present in nucleic acids. These are aromatic bases and either monocyclic pyrimidines or bicyclic purines. Ribose (RNA) and 2-deoxyribose (2-DR) are the two sugars.

2.16.1. The Double Helix Structure

In DNA, hydrogen interactions between the bases hold each polymeric strand of DNA together as it forms a helix. Bases are connected in a precise way, such as adenine links with thymine (A-T) and guanine links with cytosine (G-C).

One pyrimidine (T or C) is specifically bound to each purine (A or G) by double or triple hydrogen bonds. There are two types of hydrogen bonds: one connects an amine to a carbonyl group, and the other connects an amine to an imine (Crick, 1974).

KEYWORD

Hydrogen bonds are present abundantly in the secondary structure of proteins, and also sparingly in tertiary conformation.

2.16.2. Modified Nucleosides and Bases

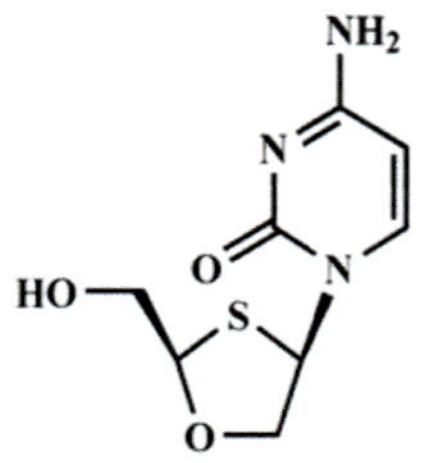

Lamivudine
Anti-AIDS drug

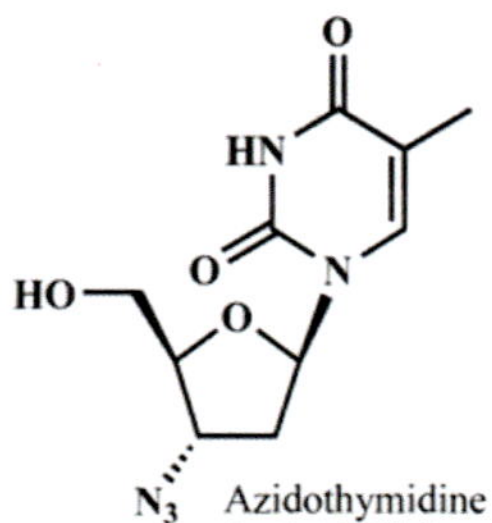

Azidothymidine
Anti-AIDS drug

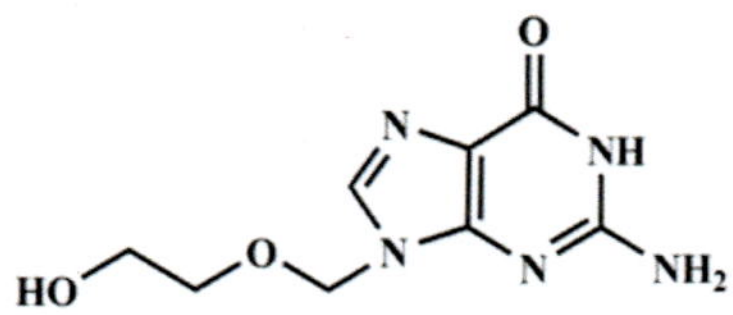

Acyclovir
Anti-herpes drug

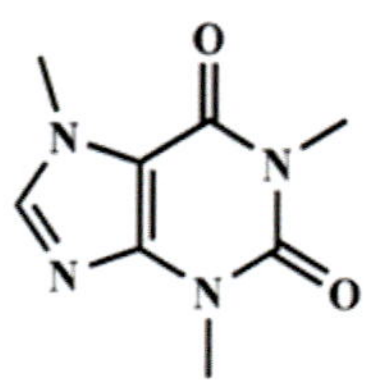

Caffeine

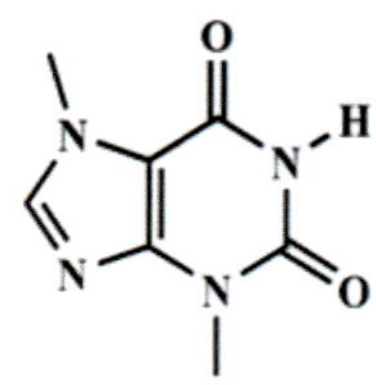

Theobromine

SUMMARY

This chapter on "Biomolecules" provides an overview of the importance of carbohydrates, proteins, lipids, and nucleic acids in the human body. Carbohydrates are an essential source of energy for the body and are found in foods such as fruits, vegetables, grains, and sugars. They also play a role in maintaining proper digestive function.

Proteins are crucial for building and repairing tissues in the body, including muscle, bone, and skin. They also help regulate various bodily functions such as metabolism, immune response, and hormone production. Lipids, including fats and oils, are important for energy storage and insulation, as well as providing protection for organs. They also serve as building blocks for hormones and cell membranes.

Nucleic acids, such as DNA and RNA, are responsible for storing and transmitting genetic information. They are essential for cell division and replication and play a critical role in protein synthesis.

Overall, understanding the roles and importance of these biomolecules can help individuals make informed decisions about their diet and overall health.

REVIEW QUESTIONS

1. What is the importance of carbohydrates in our diet? Explain with examples of some common carbohydrates and their functions in the body.
2. Proteins are essential biomolecules with a wide range of functions. Discuss the significance of proteins in the human body, including their structural, enzymatic, and regulatory roles.
3. Lipids are a diverse class of biomolecules with a variety of functions in the body. Explain the concept of lipids and their importance, including examples of different types of lipids and their functions.
4. What are nucleic acids, and what are their roles in the human body? Discuss the functions of DNA and RNA, including their roles in protein synthesis and genetic information transfer.
5. How do imbalances in the intake or metabolism of biomolecules like carbohydrates, proteins, and lipids affect human health? Discuss some common metabolic disorders related to these biomolecules and their impact on the body.

MULTIPLE CHOICE QUESTIONS

1. **Which of the following is not a function of carbohydrates in the human body?**
   a. Providing energy
   b. Regulating blood glucose levels
   c. Building and repairing tissues
   d. Enhancing immune system function
2. **Which of the following is not a type of protein?**
   a. Enzymes
   b. Hormones
   c. Antibodies
   d. Fats
3. **Which of the following is not a function of lipids in the human body?**
   a. Providing energy
   b. Insulating and protecting organs
   c. Regulating blood glucose levels
   d .Forming cell membranes
4. **Which of the following is a complex carbohydrate?**
   a. Glucose
   b. Fructose
   c. Sucrose
   d. Starch
5. **Which of the following is not an essential amino acid?**
   a. Lysine
   b. Phenylalanine
   c. Alanine
   d. Valine
6. **Which of the following is not a function of proteins in the human body?**
   a. Building and repairing tissues
   b. Regulating body functions
   c. Providing energy
   d. Fighting off infections
7. **Which of the following is not a type of lipid?**
   a. Triglyceride

b. Phospholipid
c. Cholesterol
d. Glucose

8. Which of the following is not a nucleotide base found in DNA?
a. Adenine
b. Thymine
c. Cytosine
d. Uracil

Answers to Multiple Choice Questions

1. (d)	2. (d)	3. (c)	4. (d)	5. (c)
6. (c)	7. (d)	8. (d)		

REFERENCES

1. Aggarwal, S., Tandon, C. D., Forouzandeh, M., Singla, S. K., Kiran, R., & Jethi, R. K. (2000). Role of biomolecules from human renal stone matrix on COM crystal growth. *Molecular and Cellular Biochemistry*, *210*, 109-119.
2. Albersheim, P., & Anderson-Prouty, A. J. (1975). Carbohydrates, proteins, cell surfaces, and the biochemistry of pathogenesis. *Annual Review of Plant Physiology*, *26*(1), 31-52.
3. Anson, M. L., & Mirsky, A. E. (1925). On some general properties of proteins. *The Journal of general physiology*, *9*(2), 169.
4. Bao, G. (2002). Mechanics of biomolecules. *Journal of the Mechanics and Physics of Solids*, *50*(11), 2237-2274.
5. Bergström, J. P., & Helander, A. (2008). Clinical characteristics of carbohydrate-deficient transferrin (% disialotransferrin) measured by HPLC: sensitivity, specificity, gender effects, and relationship with other alcohol biomarkers. *Alcohol & Alcoholism*, *43*(4), 436-441.
6. Brown, M. S., Kovanen, P. T., & Goldstein, J. L. (1981). Regulation of plasma cholesterol by lipoprotein receptors. *Science*, *212*(4495), 628-635.
7. Bucior, I., & Burger, M. M. (2004). Carbohydrate–carbohydrate interactions in cell recognition. *Current opinion in structural biology*, *14*(5), 631-637.
8. Callegarin, F., Quezada Gallo, J. A., Debeaufort, F., & Voilley, A. (1997). Lipids and biopackaging. *Journal of the American Oil Chemists' Society*, *74*, 1183-1192.
9. Ciucanu, I., & Kerek, F. (1984). A simple and rapid method for the permethylation of carbohydrates. *Carbohydrate Research*, *131*(2), 209-217.

10. Corriou, J. P., Iordache, O., & Tondeur, D. (1991). Classification of biomolecules by information entropy. *Journal de chimie physique*, *88*, 2645-2652.
11. Cozzone, A. J. (2002). Proteins: Fundamental chemical properties. *Encyclopedia of Life Sciences. No month listed—2002. John Wiley & Sons Ltd, 1*(1), 1-10.
12. Crick, F. (1974). The double helix: a personal view. *Nature*, *248*(5451), 766-769.
13. Crippen, G. M. (1978). The tree structural organization of proteins. *Journal of molecular biology*, *126*(3), 315-332.
14. Cronan Jr, J. E., & Gelmann, E. P. (1975). Physical properties of membrane lipids: biological relevance and regulation. *Bacteriological Reviews*, *39*(3), 232-256.
15. Cummings, J. H., & Stephen, A. M. (2007). Carbohydrate terminology and classification. *European journal of clinical nutrition*, *61*(1), S5-S18.
16. Di Pasquale, M. G. (2009). The essentials of essential fatty acids. *Journal of dietary supplements*, *6*(2), 143-161.
17. Du, Z., Hu, B., Ma, X., Cheng, Y., Liu, Y., Lin, X., ... & Ruan, R. (2013). Catalytic pyrolysis of microalgae and their three major components: carbohydrates, proteins, and lipids. *Bioresource Technology*, *130*, 777-782.
18. Fahy, E., Cotter, D., Sud, M., & Subramaniam, S. (2011). Lipid classification, structures and tools. *Biochimica et Biophysica Acta (BBA)-Molecular and Cell Biology of Lipids*, *1811*(11), 637-647.
19. Fleck, A., & Munro, H. N. (1966). The determination of nucleic acids. *Methods of biochemical analysis*, *14*, 113-176.
20. Foegeding, E. A., Luck, P. J., & Davis, J. P. (2006). Factors determining the physical properties of protein foams. *Food hydrocolloids*, *20*(2-3), 284-292.
21. Hackermüller, L., Uttenthaler, S., Hornberger, K., Reiger, E., Brezger, B., Zeilinger, A., & Arndt, M. (2003). Wave nature of biomolecules and fluorofullerenes. *Physical Review Letters*, *91*(9), 090408.
22. Hamilton, J. A. (2007). New insights into the roles of proteins and lipids in membrane transport of fatty acids. *Prostaglandins, leukotrienes and essential fatty acids*, *77*(5-6), 355-361.
23. Hepler, J. R., & Gilman, A. G. (1992). G proteins. *Trends in biochemical sciences*, *17*(10), 383-387.
24. Hutchison, W. C., & Munro, H. N. (1961). The determination of nucleic acids in biological materials. A review. *Analyst*, *86*(1029), 768-813.
25. Jakobek, L. (2015). Interactions of polyphenols with carbohydrates, lipids and proteins. *Food Chemistry*, *175*, 556-567.
26. Janmey, P. A., & Kinnunen, P. K. (2006). Biophysical properties of lipids and dynamic membranes. *Trends in cell biology*, *16*(10), 538-546.
27. Jequier, E. (1994). Carbohydrates as a source of energy. *The American Journal of clinical nutrition*, *59*(3), 682S-685S.

28. Karplus, M., & McCammon, J. A. (2002). Molecular dynamics simulations of biomolecules. *Nature structural biology*, *9*(9), 646-652.

29. Killian, J. A. (1998). Hydrophobic mismatch between proteins and lipids in membranes. *Biochimica et Biophysica Acta (BBA)-Reviews on Biomembranes*, *1376*(3), 401-416.

30. Kimura, K. (1996). Role of essential trace elements in the disturbance of carbohydrate metabolism. *Nihon rinsho. Japanese Journal of Clinical Medicine*, *54*(1), 79-84.

31. Kinsella, J. E., & Melachouris, N. (1976). Functional properties of proteins in foods: a survey. *Critical Reviews in Food Science & Nutrition*, *7*(3), 219-280.

32. Knowles, P. F., Marsh, D., & Rattle, H. W. E. (1976). Magnetic resonance of biomolecules. *An Introduction to the Theory and Practice of NMR and ESR in Biological Systems, 1*(1), 1-207.

33. Lee, Y. C. (1992). Biochemistry of carbohydrate-protein interaction 1. *The FASEB Journal*, *6*(13), 3193-3200.

34. Li, J., Yuan, X., Dong, Z., Mugabe, W., & Shao, T. (2018). The effects of fibrolytic enzymes, cellulolytic fungi and bacteria on the fermentation characteristics, structural carbohydrates degradation, and enzymatic conversion yields of *Pennisetum sinese* silage. *Bioresource Technology*, *264*, 123-130.

35. Liu, Y., Beyer, A., & Aebersold, R. (2016). On the dependency of cellular protein levels on mRNA abundance. *Cell*, *165*(3), 535-550.

36. Marsh, J. B., & Weinstein, D. B. (1966). Simple charring method for determination of lipids. *Journal of lipid research*, *7*(4), 574-576.

37. McGandy, R. B., Hall, B., Ford, C., & Stare, F. J. (1972). Dietary regulation of blood cholesterol in adolescent males: a pilot study. *The American Journal of Clinical Nutrition*, *25*(1), 61-66.

38. McNaught, A. D. (1997). Nomenclature of carbohydrates. *Carbohydrate Research*, *297*(1), 1-92.

39. Muro, E., Atilla-Gokcumen, G. E., & Eggert, U. S. (2014). Lipids in cell biology: how can we understand them better?. *Molecular biology of the cell*, *25*(12), 1819-1823.

40. Noike, T., Endo, G., Chang, J. E., Yaguchi, J. I., & Matsumoto, J. I. (1985). Characteristics of carbohydrate degradation and the rate-limiting step in anaerobic digestion. *Biotechnology and Bioengineering*, *27*(10), 1482-1489.

41. O'Keefe, S. F. (2002). Nomenclature and classification of lipids. In *Food Lipids, CRC Press*, *1*(1), (pp. 20-59).

42. Park, S., & Shin, I. (2002). Fabrication of carbohydrate chips for studying protein–carbohydrate interactions. *Angewandte Chemie*, *114*(17), 3312-3314.

43. Sargent, J. R., Tocher, D. R., & Bell, J. G. (2003). The lipids. *Fish Nutrition,1* (1), 181-257.

44. Sharon, N., & Lis, H. (1993). Carbohydrates in cell recognition. *Scientific American*, *268*(1), 82-89.

45. Somero, G. N. (1995). Proteins and temperature. *Annual Review of Physiology*, *57*(1), 43-68.
46. Su, D. S. (2011). Inorganic Materials with Double-Helix Structures. *Angewandte Chemie International Edition*, *50*(21), 4747-4750.
47. Thompson, M. R., & Thorpe, T. A. (1987). Metabolic and non-metabolic roles of carbohydrates. *Cell and tissue culture in forestry: general principles and biotechnology, 1*(1), 89-112.
48. Van Meer, G., Voelker, D. R., & Feigenson, G. W. (2008). Membrane lipids: where they are and how they behave. *Nature Reviews Molecular cell biology*, *9*(2), 112-124.
49. Wen, P. C., Mahinthichaichan, P., Trebesch, N., Jiang, T., Zhao, Z., Shinn, E., ... & Tajkhorshid, E. (2018). Microscopic view of lipids and their diverse biological functions. *Current opinion in structural biology*, *51*, 177-186.
50. Werz, D. B., & Seeberger, P. H. (2005). Carbohydrates as the next frontier in pharmaceutical research. *Chemistry–A European Journal*, *11*(11), 3194-3206.
51. Weymouth-Wilson, A. C. (1997). The role of carbohydrates in biologically active natural products. *Natural Product Reports*, *14*(2), 99-110.
52. Wihodo, M., & Moraru, C. I. (2013). Physical and chemical methods used to enhance the structure and mechanical properties of protein films: A review. *Journal of Food engineering*, *114*(3), 292-302.
53. Yue, K., & Dill, K. A. (1995). Forces of tertiary structural organization in globular proteins. *Proceedings of the National Academy of Sciences*, *92*(1), 146-150.

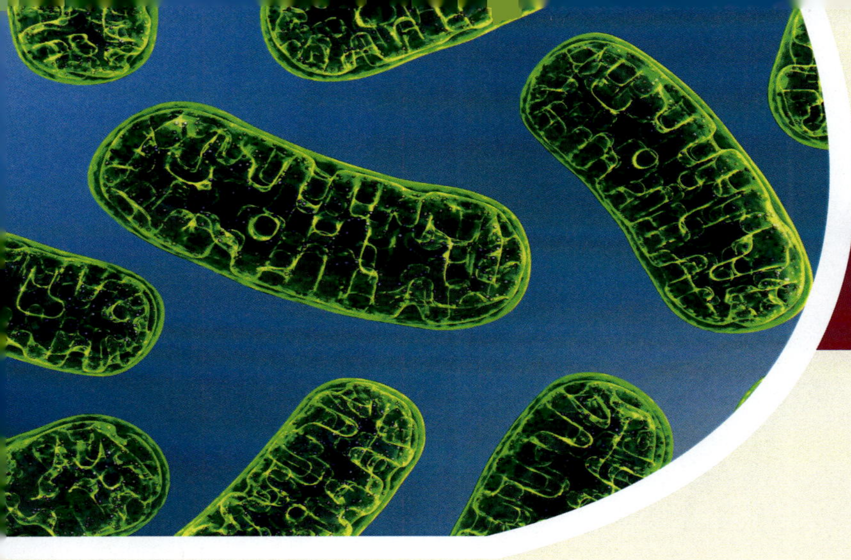

CHAPTER 3

ENERGY AND METABOLISM

UNIT INTRODUCTION

Bioenergetics is a term used by researchers to describe the flow of energy between different parts of a living organism. Periodic and sequential chemical reactions underlie cellular processes like the formation and dissolution of complex molecules (Taegtmeyer et al., 1998). There are two main types of chemical reactions: those that take no external energy to take place (called "spontaneous") and those that do (called "non-spontaneous") (Gollnick, 1985; Stanley et al., 1997). In the same way that organisms must constantly take in nutrients to maintain their energy levels, cells within living organisms must constantly generate new energy to replace that which is consumed in the process of chemical reactions (Bottomley, 1985; Jacobus, 1985; Welch et al., 1989). All the chemical reactions that happen in cells happen there. Metabolism refers to the chemical processes within a cell that either produce or use energy (Boothby et al., 1936; Neubauer et al., 1992; Flores et al., 2000).

TIPS

Keep in mind that metabolism is not just about burning calories, but also about how efficiently your body converts food into energy. To support a healthy metabolism, focus on eating a balanced diet, staying hydrated, and engaging in regular physical activity.

Figure 3.1. Illustration of energy capturing from the sun (Source: Petraru et al., Creative Commons License).

Learning Objectives

At the end of this chapter, the reader will be able to understand

- The metabolic pathways of the human body
- The importance of energy in the human body
- The concept of enzymes and their importance
- The significance of feedback inhibition in metabolic paths

Key Terms

- Photosynthetic Activity
- Anabolism
- Thermodynamics
- Entropy
- Endergonic Processes
- Exergonic Processes
- Hydrophilic
- Hydrophobic

3.1. METABOLIC PATHWAYS

A prominent example of a biological function that both consumes and creates energy is the metabolism of sugar. Because sugar molecules have a significant amount of energy trapped inside their chemical bonds, living species use sugar as a primary energy source (Tolbert, 1981). The majority of the time, these sugar molecules are produced by photosynthesizing organisms like plants (Reddy et al., 1993). Plants use energy from sunlight to convert atmospheric carbon dioxide (CO_2) into sugar molecules during the photosynthetic activity (e.g. $C_6H_{12}O_6$). They take in CO_2, and as a byproduct, they produce oxygen. This response is demonstrated by:

$6CO_2+6H_2O$ --> $C6H_{12}O6+6O_2$

KEYWORD

Electrical energy is the energy derived from electric potential energy or kinetic energy of the charged particles. In general, it is referred to as the energy that has been converted from electric potential energy.

Some quantity of energy is vital for carrying out the aforementioned reaction since it leads to the creation of the chemical that stores energy. Adenosine triphosphate (ATP), which is regarded as the main energy source of all cells, provides the majority of the energy needed for photosynthetic reactions. Contrarily, energy-storing molecules (like glucose) are only used to produce energy when bonds are broken (Keller & Hohn, 1997; Schuster et al., 2000). Oxygen is necessary for the reaction that produces energy from a sugar molecule found in cells. The photosynthesis reaction can be reversed to produce this reaction. In respration, oxygen is used up, and as a byproduct, CO_2 is generated in photosynthesis (Boroughs & DeBerardinis, 2015). Below is the equation:

Respiration: $C6H_{12}O6+6O_2$ --> $6CO_2+6H_2O$

Photosynthesis: 6CO2 + 6H2O —> C6H12O6 + 6O2

Both of the aforementioned reactions require several steps to complete. Two examples of metabolic pathways are shown by the processes used to build and break down sugar molecules. A set of chemical events known as "metabolic pathways" alter a beginning molecule in steps through a series of metabolic intermediates to produce a finished product (Zaslaver et al., 2004). In the example of sugar metabolism, the first metabolic pathway resulted in the production of sugar from smaller molecules of water and carbon dioxide, whereas the second pathway caused the decomposition of sugar into simpler compounds (E.L. Pearce & E.J. Pearce, 2013). These two distinct processes, the first of which requires energy and the second of which produces energy, are referred to as anabolic routes (i.e., the synthesis of polymers) and catabolic pathways, respectively. Hence, synthesis (also known as anabolism) and breakdown make up metabolism (i.e. catabolism). It is crucial to realize that metabolic pathways contain chemical processes that do not happen on their own. An enzyme is a protein that speeds up (catalyzes) each step of the reaction. All biological reactions, both those that need energy to continue and those that release energy, depend on enzymes to be catalyzed (Koller et al., 2002; Papin et al., 2003).

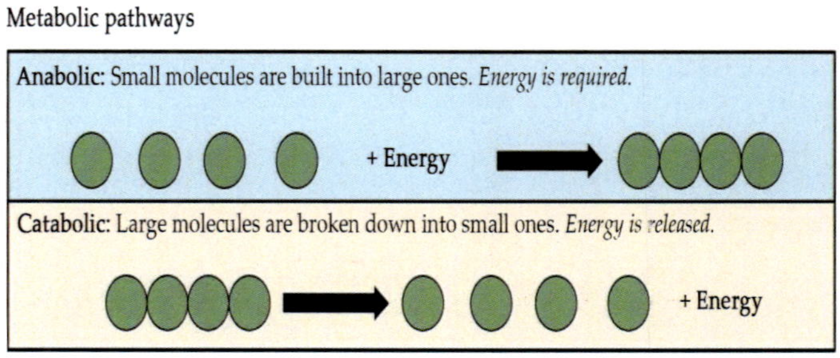

Figure 3.2. Schematic of the purpose of catabolic pathways is to produce energy by destroying bigger molecules. Anabolic pathways are made to generate bigger molecules with the help of outside energy. Both types of pathways are essential for preserving a cell's energy balance (Source: Lumen Learning, Creative Commons License).

3.2. ENERGY

Organisms in biology referred to as open systems. Through open standards and their environment, energy is exchanged. These systems engage in photosynthesis or consume molecules that store energy before releasing energy into the surroundings (Christian et al., 1994; Kornienko et al., 2016). Energy systems are subject to the same physical rules that apply to all other physical existence in the physical universe. The transmission of energy inside and between each system in the universe is governed by the principles of thermodynamics. Energy is the capacity to do work or to bring about any kind of transformation (Boyes & Stanisstreet, 1991; Muller, 1995). Energy can take many different forms. For instance, there are various types of energy, such as electrical energy, heat energy, and light energy. Understanding two physical thermodynamic laws—which govern how energy flows into and out of various biological systems—is crucial for illustrating how energy flows in and out of these systems. (Momirlan & Veziroglu, 2005; Wolpert, 2016).

3.2.1. Thermodynamics

The universe's overall sum of all energy is constant and always conserved, according to the first thermodynamic law. It asserts that the universe's overall energy level has remained constant and will do so indefinitely (Katchalsky & Kedem, 1962; Toussaint & Schneider, 1998). There are many distinct types of energy in the universe.

The first thermodynamic law provides that although energy can be transformed into different forms or transported from one place to another, it cannot be synthesized or destroyed. Around us, energy is constantly being transferred and transformed. Lighting fixtures, for instance, transform electrical energy into thermal and light energy. Natural gas's chemical energy is transformed into heat energy by gas stoves. The conversion of solar energy into chemical energy, which is stored in organic molecules, is one of the most crucial biological energy conversions that plants perform.

Did you Know?

The second law of thermodynamics states that the total entropy of an isolated system will always tend to increase over time, meaning that systems tend towards disorder rather than order.

All living systems face difficulties in obtaining energy from the surroundings in forms that they can use or transform into usable energy for work. To address this energy constraint, living cells have undergone a phase of evolution. Organic compounds often store chemical energy (e.g. sugars and fats). This energy is transported and transformed into energy inside ATP molecules via a number of chemical processes that take place at the cellular level. The energy found in ATP molecules is easily usable for work. The cells carry materials, build complex compounds, power the movements of flagella or cilia, and stretch muscle fibers to create movement as examples of the types of function they conduct (Gladyshev, 1978; 2002).

It may appear simple that a living cell's main task is to acquire, transform, and consume energy in order to perform a function. Yet, the second thermodynamic law explains why these tasks are more difficult than they appear. Energy is always transformed and transferred in a partially efficient manner. Every energy transmission results in some energy being lost in an unusable form. The majority of the time, this energy manifests as heat. According to thermodynamics, heat energy is any energy that is transferred from one physical process to another but does not take the form of work. For instance, a certain amount of electrical energy that is transformed into light energy when we put on a light bulb is lost from the systems as heat energy. Similar to this, some energy escapes as heat energy in cellular-scale metabolic activities.

The concept of ordering and disorder in a physical system is crucial. A system is more random and disorganized the more energy it loses to its environment. Entropy of the system is the term used by scientists to describe this level of disorder or unpredictability in a system. Low energy and significant disorder are indicators of strong entropy. Entropy is a changeable property of molecules and chemical reactions. For instance, as molecules that exist in high concentrations at one location diffuse out and spread, the system's overall entropy increases. According to the second rule of thermodynamics, heat energy will always be lost through energy transmissions or transformations.

Living things are highly organized species that need a constant energy intake to maintain a low entropy state.

Figure 3.3. Examples of energy transmission and conversion from one body to another as well as between different states are displayed. Similar to how light energy enables plants to produce the chemical energy they require, the food we eat gives our bodies the energy they need to carry out biological processes (Source: Lumen Learning, Creative Commons License).

3.2.2. Kinetic and Potential Energy

An energy that is driving the motion of an object is present. For instance, a wrecking ball in slow motion can significantly harm anything in its path (Turvey et al., 1999). Kinetic energy is the name given to the energy present in moving objects. A human walking, a bullet traveling at high speed, and air molecules moving quickly (which produce heat) all have kinetic energy (Levitt et al., 1995).

A moving ball will still contain energy if lifted to a certain height while remaining immobile because of changes in the potential difference between height and gravity. We gave the ball a certain amount of energy when we raised it. Due to its positioning and the gravitational force pulling on it, the ball has now stored that energy. Potential energy is the name for this form of energy. The accumulated potential energy would be turned into kinetic energy (during movement) if the identical ball were to fall from that height, and this process would continue until all of the ball's potential energy was used up and it came to rest on the earth. Wrecking ball has a swing that resembles a pendulum. Due to changes in position and speed, the swing generates variable energy levels (both potential and kinetic). Whereas kinetic energy is maximum at the swing bottom, potential energy is most at the top of the swing. Some examples of potential energy are the energy present in water held behind such a dam or the energy a person has just before skydiving out of an airplane (Odum & Pinkerton, 1955).

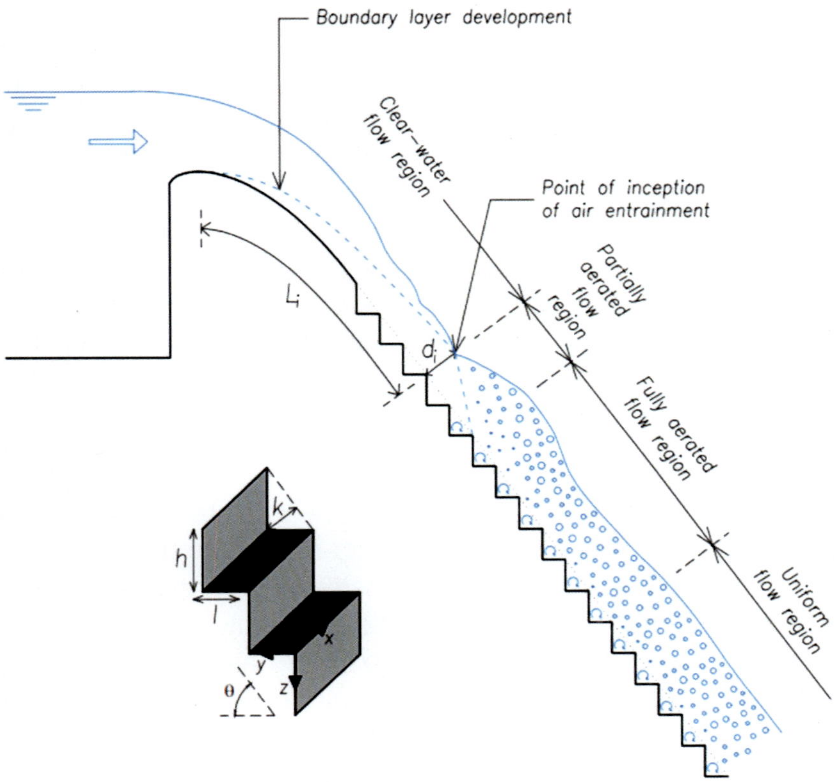

Figure 3.4. Image of water that is still and contains potential energy, while water that is moving (like a cascade or a river that is flowing) includes kinetic energy (Source: Matos et al., Creative Commons License).

In addition to an object's location, potential energy also depends on the framework of the thing. For instance, if we compress a spring that is on the ground, it has potential energy. A rubber band that has been drawn taut also has elastic energy in it. At the molecular level, the bonds that bind the atoms of the molecules together have a specific structure that holds the potential energy. Keep in mind the idea that anabolic biological pathways require energy to generate complex structures from small (simpler) molecules, whereas catabolic pathways emission owing to the breakdown of complex molecules (Smith et al., 1980; Adair, 1991). The fact that energy is released when bonds break suggests that these bonds have some stored potential energy. It is a known truth that every consumable we consume contains some potential energy (retained in their bonds), which is used by us once the bonds have been broken. Only when these bonds are broken do they release energy. Chemical energy refers to the type of potential energy that is contained inside chemical bonds and released when those connections are broken. The food molecules give the living cells chemical energy (Andrec et al., 2005).

KEYWORD

Potential energy, stored energy that depends upon the relative position of various parts of a system. A spring has more potential energy when it is compressed or stretched.

3.2.3. Free and Activation Energy

It is critical to measure this energy after knowing that chemical reactions release a certain level of energy as a result of the dissolution of energy-storing bonds. The energies associated with a chemical process are measured using the free energy related to a system. Keep in mind that according to the second rule of thermodynamics, all energy transmissions include the loss of some energy (heat) in a useless form. Free energy refers to the quantity of energy associated with a chemical process that is still within the system after accounting for energy losses. Free energy is essentially any energy that can be used or obtained to carry out tasks (Eisenberg & Hill, 1985).

If energy is released during a chemical process, the free energy change, written as delta G or ΔG, will be negative. A negative ΔG also denotes that a reaction's products have less free energy than its reactants do. Reactants in this instance emit extra free energy as the reaction progresses. Exergonic reactions are those that have a negative ΔG Exergonic means positive energy is escaping the system. The reactants of these exergonic (or exothermic) reactions, also known as spontaneous reactions, have more stored energy than the products. The phrase "spontaneous reaction" and the statement that a chemical reaction is occurring right away have a significant difference. Contrary to common usage, spontaneous reactions don't happen instantly or fast; instead, they take their time and don't consume any energy to move forward. A perfect example of a slow-moving spontaneous reaction is the rusting of iron (but lacking the usage of outside energy) (Kötting & Gerwert, 2004).

The increase in free energy for a certain chemical process will be positive (i.e., +G) if a positive energy is absorbed rather than released. In such situations, reactants typically have less energy than products. As a result, the end products of these processes might be thought of as compounds that store energy (molecules). Non-spontaneous endergonic reactions are the name given to these chemical processes. Even without the inclusion of outside free energy, an endothermic reaction (or endothermic) phenomenon is not possible (Jortner, 1976; Tomasi et al., 1980).

KEYWORD

Catalyst, in chemistry, any substance that increases the rate of a reaction without itself being consumed. Enzymes are naturally occurring catalysts responsible for many essential biochemical reactions.

Figure 3.5. Examples of both endergonic and exergonic processes (those that demand energy) (ones which discharge energy) (Source: Open text bc, Creative Commons License).

Whether working with either endothermic or exothermic reactions, there are a few basic ideas that must be recalled. Before beginning the energy-releasing phases, exothermic reactions require a little quantity of input energy to start the reaction. As a result, these reactions produce a net energy release but require initial input energy. The activation energy is the small amount of energy input required for all biological reactions to occur (Cramer & Truhlar, 1992; Krasnovsky et al., 2003).

3.3. ENZYMES

The molecules that are utilized in the catalytic of biochemical processes are called as enzymes. A catalyst is a material that helps a biochemical reaction to occur. The majority of enzymes are proteins that play crucial roles in lowering the activation energy of chemical processes occurring inside cells (Boyer & Krebs, 1986). The majority of crucial metabolic processes occur at optimum temperatures far too slowly for the benefit of the cells. Without enzymes to accelerate the processes, life could not exist. Enzymes attach to the components of the reaction mixture and retain them in a useful manner so that chemical bond-forming and bond-breaking reactions can take place more quickly. It is crucial to keep in mind that enzymes have the same characteristics whether a reaction is endothermic or exothermic (spontaneous).

Remember

Enzymes are incredibly powerful catalysts that can speed up chemical reactions by a factor of millions or even billions. However, they are also incredibly specific and can only catalyze certain reactions under specific conditions. Understanding enzyme structure and function is crucial for developing new drugs and therapies.

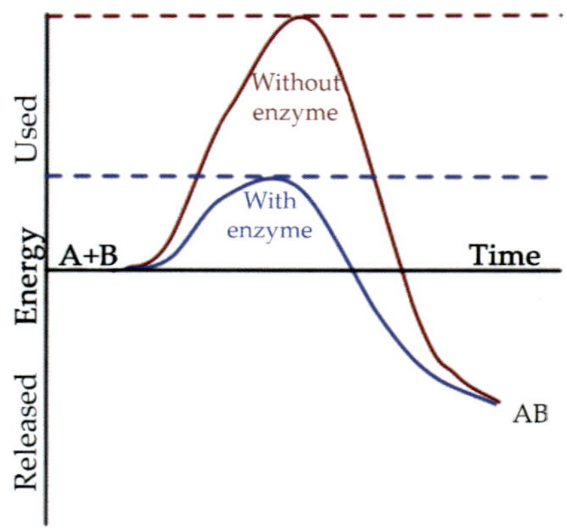

Figure 3.6. Schematic of enzymes typically reduce a reaction's activation energy while leaving the reaction's free energy unaffected (Source: wikimedia, Creative Commons License).

This is because they don't change the reactants' or products' free energies. They only result in a reduction in the reaction's activation energy. Moreover, the reaction that is being catalyzed leaves an enzyme intact. The enzyme is ready for use in further reactions after catalyzing one (Krebs & Beavo, 1979; Murakami et al., 1996).

Chemical reactants are the substrates, to which enzymes connect. Depending on the specific chemical reactions, there may be one or more substrates. A single reactant, or substrate, can split up into many products in various chemical processes. On the other hand, two substrates can combine to form a single, bigger molecule. Two reactant substrates, however, can potentially produce two products with different features. The active site of an enzyme is the region within the enzyme where substrates often bind. As enzymes are fundamentally proteins, their active sites have a unique arrangement of side amino acid chains. There are various subcategories for each side chain. They come in many shapes and sizes, and might be neutral, basic, slightly acidic, hydrophilic, or hydrophobic (Lindahl, 1982).

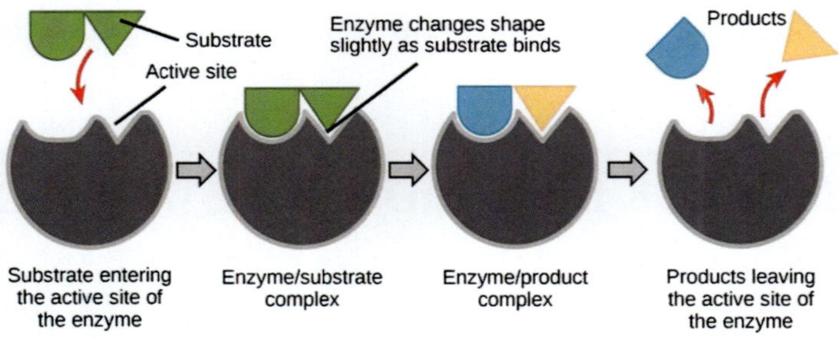

Figure 3.7. The lock-and-key model, which explains how enzymes and substrates dynamically change throughout the transition state to increase the substrates' affinity for the active sites, is modified by the induced-fit model (Source: wikimedia, Creative Commons License).

Enzyme activity in some cell settings is regulated by activation energy as well as external elements like temperature, pH, salt content, etc. (Rawlings & Salvesen, 2013). Enzyme activity is also managed in ways that might either promote or prevent it. Several types of chemicals can inhibit or enhance the activity of enzymes. In some instances of inhibition, the effect of specific binds to the active site of the enzyme in a manner that is extremely similar to the substrate, preventing the substrate molecule from attaching. Owing to this inhibition, competitive inhibition is used to block the enzyme. This is because a substrate molecule and an inhibitor molecule compete to bind to an active site.

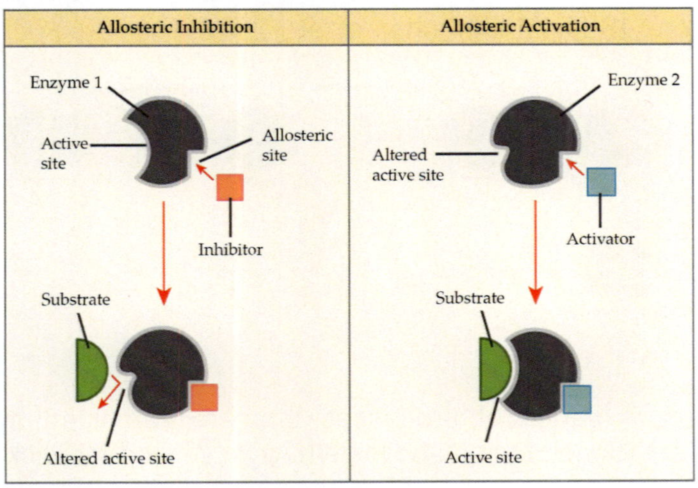

Figure 3.8. Allosteric inhibition works by subtly inducing a conformational change to an active site that prevents the substrates from fitting. On the other hand, in allosteric activation, the activator molecule modifies the geometry of the active site to provide a better match for the substrate (Source: Philschatz, Creative Commons License).

On the other hand, non-competitive inhibition occurs when an inhibitor molecule binds to the enzyme at an allosteric site, which is a non-active location. In this instance, the enzyme is nonetheless successful in preventing substrate binding to the active site. A few of the molecules bind to enzymes at sites where their fusion causes a conformational change that lessens the enzyme's attraction to the substrate. **Allosteric inhibition** is the name given to this type of inhibition. There are also allosteric activators in addition to inhibitors. The conformation of the enzyme changes as a result of the attachment of allosteric activators to regions of the non-active site, increasing the affinity of the enzyme's active site for the substrate.

ACTIVITY 3.1.

A fitness company is launching a new energy drink that claims to boost metabolism and increase fat burning. How does metabolism work in the body, and what scientific evidence supports the effectiveness of such products?

3.4. FEEDBACK INHIBITION IN METABOLIC PATHS

There are numerous ways that molecules can control how an enzyme works (Fell, 1992; 1998; Teusink et al., 1998). The end products of metabolic reactions that take place at the cellular level are the most important sources of governing molecules (with regard to metabolic reactions of enzymes). Cells have evolved to use the outputs generated by their own reactions to suppress the feedback of enzyme activity in the most elegant and effective way possible. A feedback inhibition system uses a reaction yield to regulate its own subsequent production (Simao et al., 2005). The cell slows down synthesis during catabolism or anabolic responses in response to an abundance of the products. By using the methods and techniques outlined above, these types of byproducts can prevent the enzymes that catalyzed their creation from doing their job (Sheen, 1994; Mulhbacher et al., 2010).

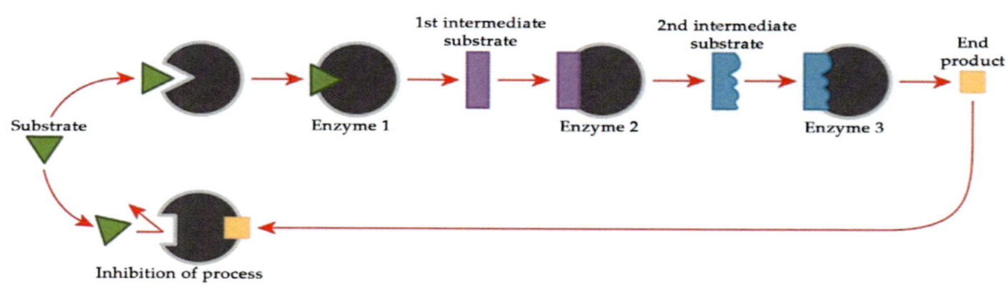

Figure 3.9. A series of processes that are catalyzed by several enzymes are formed by metabolic pathways. When feedback inhibition occurs, an upstream process is inhibited by the pathway's end product (a crucial regulatory mechanism occurring in cells) (Source: Philschatz, Creative Commons License).

SUMMARY

This chapter "Energy and Metabolism" discusses the metabolic pathways of the human body, the importance of energy in the body, the concept of enzymes and their importance, and the significance of feedback inhibition in metabolic pathways. Metabolism is the set of chemical reactions that occur in living organisms to maintain life. The chapter explains how these reactions occur through metabolic pathways, which involve a series of enzyme-catalyzed reactions. Energy is required for these reactions to occur, and the chapter describes how the body obtains and utilizes energy from various sources such as carbohydrates, fats, and proteins. Enzymes play a crucial role in these pathways by catalyzing reactions and regulating metabolic processes. The chapter also discusses feedback inhibition, which is a process that regulates metabolic pathways by inhibiting enzyme activity. Understanding these concepts is essential to comprehend the complex biochemical processes that occur in the human body.

REVIEW QUESTIONS

1. What are the major metabolic pathways in the human body and what are their functions?
2. How does the human body acquire and use energy, and what are the consequences of energy imbalances?
3. What are enzymes, and how do they facilitate metabolic reactions in the human body?
4. What is feedback inhibition, and how does it regulate metabolic pathways in the human body?
5. How do the structure and function of enzymes contribute to their specificity and efficiency in metabolic reactions?

MULTIPLE CHOICE QUESTIONS

1. Which of the following is NOT one of the three main stages of cellular respiration?

a. Glycolysis
b. Krebs cycle
c. Electron transport chain
d. Fermentation

2. Which type of energy is stored in chemical bonds and is released during cellular respiration?

a. Kinetic energy
b. Potential energy

c. Radiant energy
d. Thermal energy

3. Which of the following is the primary source of energy for most cells?

a. Carbohydrates
b. Proteins
c. Fats
d. Minerals

4. Which of the following is an example of an anabolic pathway?

a. Glycolysis
b. Krebs cycle
c. Photosynthesis
d. Fermentation

5. What is the role of enzymes in metabolic pathways?

a. To absorb energy
b. To release energy
c. To speed up chemical reactions
d. To slow down chemical reactions

6. Which of the following is an example of a catabolic pathway?

a. Photosynthesis
b. Glycolysis
c. Protein synthesis
d. DNA replication

7. Which of the following is NOT true about enzymes?

a. Enzymes are proteins that catalyze biochemical reactions
b. Enzymes can increase the rate of a reaction by lowering the activation energy
c. Enzymes are consumed during the reaction and need to be replenished
d. Enzymes are specific for the substrates they bind to

Answers to Multiple Choice Questions

1. (d) 2. (b) 3. (a) 4. (c) 5. (c)
6. (b) 7. (c)

REFERENCES

1. Adair, R. K. (1991). Constraints on biological effects of weak extremely-low-frequency electromagnetic fields. *Physical Review A*, *43*(2), 1039.
2. Andrec, M., Felts, A. K., Gallicchio, E., & Levy, R. M. (2005). Protein folding pathways from replica exchange simulations and a kinetic network model. *Proceedings of the National Academy of Sciences*, *102*(19), 6801-6806.
3. Boothby, W. M., Berkson, J., & Dunn, H. L. (1936). Studies of the energy of metabolism of normal individuals: a standard for basal metabolism, with a nomogram for clinical application. *American Journal of Physiology-Legacy Content*, *116*(2), 468-484.
4. Boroughs, L. K., & DeBerardinis, R. J. (2015). Metabolic pathways promoting cancer cell survival and growth. *Nature cell biology*, *17*(4), 351.
5. Bottomley, P. A. (1985). Noninvasive study of high-energy phosphate metabolism in human heart by depth-resolved 31P NMR spectroscopy. *Science*, *229*(4715), 769-772.
6. Boyer, P. D., & Krebs, E. G. (1986). *The enzymes* (Vol. 17, pp. 1-26). Academic Press.
7. Boyes, E., & Stanisstreet, M. (1991). Misconceptions in first-year undergraduate science students about energy sources for living organisms. *Journal of Biological Education*, *25*(3), 209-213.
8. Christian, D. P., Niemi, G. J., Hanowski, J. M., & Collins, P. (1994). Perspectives on biomass energy tree plantations and changes in habitat for biological organisms. *Biomass and bioenergy*, *6*(1-2), 31-39.
9. Cramer, C. J., & Truhlar, D. G. (1992). An SCF solvation model for the hydrophobic effect and absolute free energies of aqueous solvation. *Science*, *256*(5054), 213-217.
10. Eisenberg, E., & Hill, T. L. (1985). Muscle contraction and free energy transduction in biological systems. *Science*, *227*(4690), 999-1006.
11. Fell, D. A. (1992). Metabolic control analysis: a survey of its theoretical and experimental development. *Biochemical Journal*, *286*(Pt 2), 313.
12. Fell, D. A. (1998). Increasing the flux in metabolic pathways: a metabolic control analysis perspective. *Biotechnology and Bioengineering*, *58*(2-3), 121-124.
13. Flores, C. L., Rodríguez, C., Petit, T., & Gancedo, C. (2000). Carbohydrate and energy-yielding metabolism in non-conventional yeasts. *FEMS Microbiology Reviews*, *24*(4), 507-529.
14. Gladyshev, G. P. (1978). On the thermodynamics of biological evolution. *Journal of Theoretical Biology*, *75*(4), 425-441.
15. Gladyshev, G. P. (2002). Thermodynamics of biological evolution and aging: Supramolecular thermodynamics is a key to understanding phenomena of life. What is life from a physical chemist's viewpoint. In *Electronic Journal of Mathematical and Physical Sciences: Conference/Seminar Edition* (Vol. 1, No. 2, pp. 1-15).
16. Gollnick, P. D. (1985, February). Metabolism of substrates: energy substrate metabolism

during exercise and as modified by training. In *Federation proceedings* (Vol. 44, No. 2, pp. 353-357).

17. Jacobus, W. E. (1985). Respiratory control and the integration of heart high-energy phosphate metabolism by mitochondrial creatine kinase. *Annual Review of Physiology*, *47*(1), 707-725.
18. Jortner, J. (1976). Temperature dependent activation energy for electron transfer between biological molecules. *The Journal of Chemical Physics*, *64*(12), 4860-4867.
19. Katchalsky, A., & Kedem, O. (1962). Thermodynamics of flow processes in biological systems. *Biophysical journal*, *2*(2 Pt 2), 53.
20. Keller, N. P., & Hohn, T. M. (1997). Metabolic pathway gene clusters in filamentous fungi. *Fungal Genetics and Biology*, *21*(1), 17-29.
21. Koller, A., Washburn, M. P., Lange, B. M., Andon, N. L., Deciu, C., Haynes, P. A., ... & Wolters, D. (2002). Proteomic survey of metabolic pathways in rice. *Proceedings of the National Academy of Sciences*, *99*(18), 11969-11974.
22. Kornienko, N., Sakimoto, K. K., Herlihy, D. M., Nguyen, S. C., Alivisatos, A. P., Harris, C. B., ... & Yang, P. (2016). Spectroscopic elucidation of energy transfer in hybrid inorganic–biological organisms for solar-to-chemical production. *Proceedings of the National Academy of Sciences*, *113*(42), 11750-11755.
23. Kötting, C., & Gerwert, K. (2004). Time-resolved FTIR studies provide activation free energy, activation enthalpy and activation entropy for GTPase reactions. *Chemical Physics*, *307*(2-3), 227-232.
24. Krasnovsky, A. A., Drozdova, N. N., Ivanov, A. V., & Ambartsumian, R. V. (2003). Activation of molecular oxygen by infrared laser radiation in pigment-free aerobic systems. *Biochemistry (Moscow)*, *68*(9), 963-966.
25. Krebs, E. G., & Beavo, J. A. (1979). Phosphorylation-dephosphorylation of enzymes. *Annual Review of Biochemistry*, *48*(1), 923-959.
26. Levitt, M., Hirshberg, M., Sharon, R., & Daggett, V. (1995). Potential energy function and parameters for simulations of the molecular dynamics of proteins and nucleic acids in solution. *Computer Physics Communications*, *91*(1-3), 215-231.
27. Lindahl, T. (1982). DNA repair enzymes. *Annual Review of Biochemistry*, *51*(1), 61-87.
28. Momirlan, M., & Veziroglu, T. N. (2005). The properties of hydrogen as fuel tomorrow in sustainable energy system for a cleaner planet. *International Journal of Hydrogen Energy*, *30*(7), 795-802.
29. Mulhbacher, J., Brouillette, E., Allard, M., Fortier, L. C., Malouin, F., & Lafontaine, D. A. (2010). Novel riboswitch ligand analogs as selective inhibitors of guanine-related metabolic pathways. *PLoS Pathogens*, *6*(4), e1000865.
30. Muller, A. W. (1995). Were the first organisms heat engines? A new model for biogenesis and the early evolution of biological energy conversion. *Progress in Biophysics and Molecular Biology*, *63*(2), 193-231.

31. Murakami, Y., Kikuchi, J. I., Hisaeda, Y., & Hayashida, O. (1996). Artificial enzymes. *Chemical Reviews*, *96*(2), 721-758.
32. Neubauer, S., Krahe, T., Schindler, R., Horn, M., Hillenbrand, H., Entzeroth, C., ... & Lackner, K. (1992). 31P magnetic resonance spectroscopy in dilated cardiomyopathy and coronary artery disease. Altered cardiac high-energy phosphate metabolism in heart failure. *Circulation*, *86*(6), 1810-1818.
33. Odum, H. T., & Pinkerton, R. C. (1955). Time's speed regulator: the optimum efficiency for maximum power output in physical and biological systems. *American Scientist*, *43*(2), 331-343.
34. Papin, J. A., Price, N. D., Wiback, S. J., Fell, D. A., & Palsson, B. O. (2003). Metabolic pathways in the post-genome era. *Trends in Biochemical Sciences*, *28*(5), 250-258.
35. Pearce, E. L., & Pearce, E. J. (2013). Metabolic pathways in immune cell activation and quiescence. *Immunity*, *38*(4), 633-643.
36. Rawlings, N. D., & Salvesen, G. (Eds.). (2013). *Handbook of proteolytic enzymes* (Vol. 135, p. 141). Amsterdam: Academic Press. DOI: https://doi.org/10.1016/C2009-1-60990-4.
37. Reddy, V. N., Mavrovouniotis, M. L., & Liebman, M. N. (1993, July). Petri net representations in metabolic pathways. In *ISMB* (Vol. 93, pp. 328-336).
38. Schneider, W. C., & Hogeboom, G. H. (1950). Intracellular distribution of enzymes. *J. Biol. Chem*, *183*(123), 9.
39. Schuster, S., Fell, D. A., & Dandekar, T. (2000). A general definition of metabolic pathways useful for systematic organization and analysis of complex metabolic networks. *Nature biotechnology*, *18*(3), 326.
40. Sheen, J. (1994). Feedback control of gene expression. *Photosynthesis Research*, *39*(3), 427-438.
41. Simao, E., Remy, E., Thieffry, D., & Chaouiya, C. (2005). Qualitative modelling of regulated metabolic pathways: application to the tryptophan biosynthesis in E. coli. *Bioinformatics*, *21*(suppl_2), ii190-ii196.
42. Smith, J. C., Frank, S. J., Bashford, C. L., Chance, B., & Rudkin, B. (1980). Kinetics of the association of potential-sensitive dyes with model and energy-transducing membranes: Implications for fast probe response times. *The Journal of Membrane Biology*, *54*(2), 127-139.
43. Stanley, W. C., Lopaschuk, G. D., & McCormack, J. G. (1997). Regulation of energy substrate metabolism in the diabetic heart. *Cardiovascular Research*, *34*(1), 25-33.
44. Taegtmeyer, H., King, L. M., & Jones, B. E. (1998). Energy substrate metabolism, myocardial ischemia, and targets for pharmacotherapy. *The American Journal of Cardiology*, *82*(5), 54K-60K.
45. Teusink, B., Walsh, M. C., van Dam, K., & Westerhoff, H. V. (1998). The danger of metabolic pathways with turbo design. *Trends in Biochemical Sciences*, *23*(5), 162-169.

46. Tolbert, N. E. (1981). Metabolic pathways in peroxisomes and glyoxysomes. *Annual Review of Biochemistry*, *50*(1), 133-157.

47. Tomasi, A., Albano, E., Lott, K. A. K., & Slater, T. F. (1980). Spin trapping of free radical products of CCl4 activation using pulse radiolysis and high energy radiation procedures. *FEBS Letters*, *122*(2), 303-306.

48. Toussaint, O., & Schneider, E. D. (1998). The thermodynamics and evolution of complexity in biological systems. *Comparative Biochemistry and Physiology Part A: Molecular & Integrative Physiology*, *120*(1), 3-9.

49. Turvey, M. T., Holt, K. G., LaFiandra, M. E., & Fonseca, S. T. (1999). Can the transitions to and from running and the metabolic cost of running be determined from the kinetic energy of running?. *Journal of Motor Behavior*, *31*(3), 265-278.

50. Welch, K. M. A., Levine, S. R., D'andrea, G., Schultz, L. R., & Helpern, J. A. (1989). Preliminary observations on brain energy metabolism in migraine studied by *in vivo* phosphorus 31 NMR spectroscopy. *Neurology*, *39*(4), 538-538.

51. Wolpert, D. (2016). The free energy requirements of biological organisms; implications for evolution. *Entropy*, *18*(4), 138.

52. Zaslaver, A., Mayo, A. E., Rosenberg, R., Bashkin, P., Sberro, H., Tsalyuk, M., ... & Alon, U. (2004). Just-in-time transcription program in metabolic pathways. *Nature genetics*, *36*(5), 486.

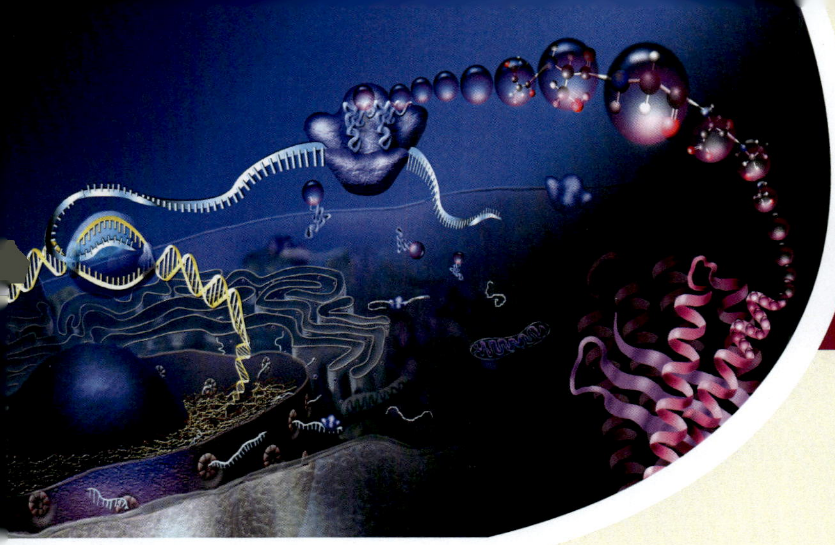

CHAPTER 4

PROTEIN SYNTHESIS AND FOLDING

UNIT INTRODUCTION

Regardless of whether a protein is found in a membrane or the extracellular space, its synthesis occurs in the cytosol, which is the inside of cells that are both prokaryotic and eukaryotic. Proteins generated by chloroplasts and mitochondria are the only ones that do not have their own set of genes to code for them. (Li et al., 2010).

Proteins are synthesized by converting the nucleotide sequences encoded on mRNA into the sequences of amino acids that compose up polypeptide chains, a process called translation. The translation of mRNA involves the binding of insignificant tRNAs to amino acids, the assembly of ribosomes (RNA& protein complexes), and the cooperation of several proteins. Some proteins regulate the interaction of these components by attaching and releasing guanosine triphosphate (GTP). The final step in the production of peptide bonds is catalyzed by ribosomal RNA molecules. Some recently produced polypeptides fold into their natural form in the cellular environment without the support of proteins known as chaperones. (Zhang et al., 2009).

All living things today use a translation process, and the technique used to synthesize peptide bonds presumably evolved from the same progenitor some 3.5 billion years ago because all modern organisms share an underlying translation mechanism. By the last common ancestor, many moderately complex governing elements had been established and transferred down the evolutionary tree (Yewdell, 2001).

Learning Objectives

By the end of this chapter, students can learn the following:

- What is protein synthesis of folding of a cell?
- What is Transfer RNA?
- Functions of ribosomes
- Protocol of protein synthesis
- What is spontaneous protein folding
- What is chaperone-assisted protein folding

Key Terms

- Protein folding
- Transfer RNA
- Protein factor
- Chaperone
- Spontaneous protein folding

4.1. MECHANISM OF PROTEIN SYNTHESIS

Nucleotides at the 3′-end regulate mRNA stability, 5′-end nucleotides provide protein binding sites, and nucleotides in the middle establish the amino acid sequencing for polypeptide synthesis. The sequence of amino acids is recognized by codons, which are groups of three nucleotides that follow one another in the protein-coding sequence. Despite a few small exceptions, the genetic code that links nucleotide triplets into amino acids is universal. Each amino acid is encoded by one to six distinct triplet codons. (Takei et al., 2004). Initiation codons (AUG) are used to initiate the formation of polypeptide chains, and methionine is always the first amino acid in a protein.

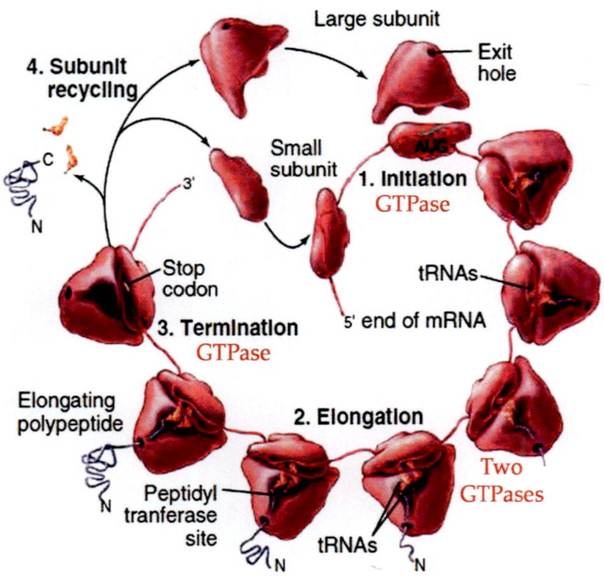

Figure 4.1. This schematic depicts the translation cycle for one mRNA strand involving six ribosomes (Source: Wikimedia, Creative Commons License).

The three ending codons can also be used to halt peptide synthesis. (UAG, UGA, UAA). There are three key features that set eukaryotic mRNAs apart from their bacterial counterparts. For starters, although eukaryotic mRNAs only translate a single protein, bacterial mRNAs often encode many proteins. Second, most eukaryotic mRNAs possess a 5′-triphosphate-5′ linkage at their 5′ ends, which is attached to an inverted remnant 7-methylguanosine. (Garcia-Esparcia, et al., 2015). This 5′ cap does not change as long as the mRNA is around. It protects the 5′ ends from nuclease attacks while providing a binding site for proteins. Third, most metazoan mRNAs require intron removal (Pawlowska et al., 2005).

Eukaryotic messenger RNAs typically have between 50 and 200 adenine residues in their 3′ tail, which is added after transcription. The poly(A) tail interacts with a protein that facilitates nuclear export & guards against cytoplasmic mRNA degradation. The 3′ poly (A) tails of bacterial mRNAs are either shortened or lacking. During translation, the double-stranded conformations of many single-stranded mRNAs must be disrupted to read each codon (Sánchez et al., 2013).

> **Did you Know?**
>
> Protein biosynthesis is inherently coupled to cotranslational protein folding. Folding of the nascent chain already occurs during synthesis and is mediated by spatial constraints imposed by the ribosomal exit tunnel as well as self-interactions.

4.2. TRNA

tRNAs act as connectors that only give amino acids to the translation process by the same mRNA codons with the proper amino acids, which are then integrated into a growing polypeptide. There are between one and four unlike tRNAs for every amino acid, and this range roughly parallels the amount of proteins that have that amino acid. Methionine-carrying tRNAs start bacterial protein synthesis (as formyl methionine). (Lowe & Eddy, 1997).

Transfer RNA comprises around 76 nucleotides, with base pairs forming four stems and three intermediate circles. When folded, these secondary structures form an L-shaped molecular shape. The "decoding" triplet or the anticodon is located at one end of the L (the acceptor's arm), and the amino acid acceptor spot is located at the terminal of the L arm. Aminoacyl-tRNA synthetases are enzymes that catalyze the first stage in a two-step process that covalently links an amino acid to its corresponding tRNA (Lowe & Chan, 2016).

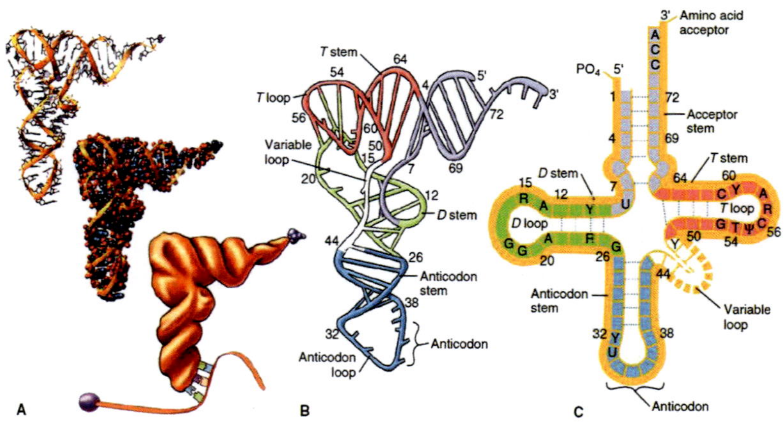

Figure 4.2. Structure of tRNA (Source: Basicmedicalkey, Creative Commons License).

Initial steps in the reaction between an amino acid and ATP result in the formation of the high-energy intermediate aminoacyl (aa) adenosine monophosphate (AMP), which releases pyrophosphate. In the next step, an aa-tRNA is formed by transferring the amino acid to the 3′ adenine of the tRNA. This step, known as "charging," generates a highly energetic ester link that links the tRNA and the amino acid, which prepares the amino acid to arrange a peptide bond with a group of amino acids in the developing polypeptide chain. Any of the 20 aa-tRNA synthetases creates an excess of tRNAs specific to a single amino acid. Two unique classes of aa-tRNA synthetases exist, distinguished by the hydroxyl groups to which the amino acids they synthesize bond at the 3' terminals of their tRNAs. (Kowalak et al., 1994).

Perfect matching of amino acids with their corresponding tRNAs is crucial for accurate protein synthesis. To make this possible, synthetases engage with the appropriate tRNA's anticodon, the tail of 3′ acceptors, and the boundary between these sites. Their proofreading processes remove incorrect amino acids on tRNAs (Lowe & Chan, 2016).

4.3. RIBOSOMES

Ribosomes are enormous macromolecular factories that use mRNA and aa-tRNAs to make polypeptides. The sequences of polypeptides generated are those that are specified by the codons sequences in the appropriate mRNAs ensured by base pairing between codons of mRNA and anticodons of tRNA. Crystal and cryoelectronic microscopy (cryo-EM) are now available for multiple ribosome types following years of effort. These structures revealed some unexpected results (Kramer et al., 2009).

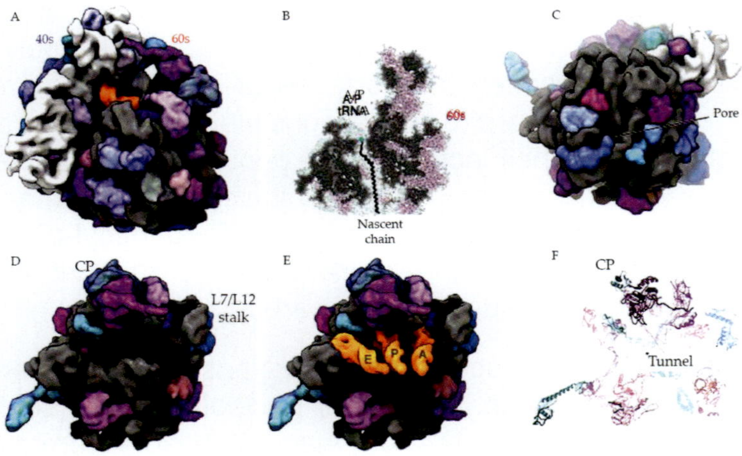

Figure 4.3. Ribosomal Structure of mammals (Source: Wikimedia Commons, Creative Commons License).

Ribosomes, which consist of a small subunit and a large subunit, assemble when an mRNA is translated. Each subunit contains many different types of proteins in addition to ribosomal RNA (rRNA) molecules. The subunit and rRNA sizes are quantified using the ultracentrifuge-determined sedimentation coefficient S (Svedburg). All ribosomes have evolved from a shared ancestor and function comparably, but their structure has altered. Mitochondrial and prokaryotic ribosomes are smaller than mammalian ribosomes, which also contain more proteins. (Zhang et al., 2009).

rRNAs form the essential basis of every ribosomal subunit. The 18S rRNA is the smallest subunit of the ribosome in mammals and consists of around 1900 nucleotides, furthermost of which are folded into base-paired helices. The RNAs that make up the subunits of the ribosome in mammals are the 28S rRNA (containing 5000 nucleotides), the 5.8S rRNA (containing 156 nucleotides), and the 5S rRNA (containing 121 nucleotides). By analyzing the sequences of rRNAs from numerous different species, it was first anticipated that the rRNAs would fold into numerous base-paired helices. These helices pack together with the loops in between to create a rigid structure (Woolhead et al., 2004).

KEYWORD

Transfer RNA is an adaptor molecule composed of RNA, typically 76 to 90 nucleotides in length, that serves as the physical link between the mRNA and the amino acid sequence of proteins.

Although prokaryotic and eukaryotic rRNAs fold similarly, the former is a distinct sequence. Some of the structures of rRNAs that have remained preserved throughout development are the subunits' interaction sides, binding regions for mRNA, tRNA, and protein cofactors, and the nucleotides in peptide bond formation. While some proteins of ribosomes are extended to peptide chains into the core of rRNA, the vast majority link to the ribosomal RNA itself. Most ribosomal proteins are simple and very small. (10–30 kD). Except in one case, every protein is only represented by a single copy on a ribosome (Waudby et al., 2019).

The interdomain space is where mRNA decoding and polypeptide synthesis occur. Because the surfaces of these areas are often protein-free, rRNAs are directly responsible for the association of mRNA, tRNA, and their activation of peptide binding. There are three spots on tRNAs that both subunits have in common: the A (aa-tRNA) site, the E (exit) site, and the P (peptidyl-tRNA) site. The larger subunit's RNA core is traversed by a tunnel via which the expanding polypeptide chain departs.

The creation and assembling of a yeast ribosome entail all known RNA polymerases, 75 smaller nucleolar RNAs (snoRNAs), over 200 protein components, more than 80 ribosomal proteins, and 4 rRNAs. Cleavage and modification of precursor RNAs provide the rRNAs. Assembly factors, which include snoRNAs and numerous proteins, guide the release of rRNAs and ribosomal proteins from the nucleus into the cytoplasm and coordinate their step-by-step assembly into the smaller and bigger subunits of the ribosome (Woodson, 2011).

Many genes encoding ribosomal proteins are essential to life, but mutations in a few can cause extremely specific disorders. Spleens are missing, but other than that, healthy people have only one copy of the gene responsible for making the ribosomal protein RPSA. Liver diseases are brought about by mutations in genes encoding specific assembly proteins, whereas anemia is triggered by abnormalities in genes encoding various ribosomal subunits (Das et al., 2008).

KEYWORD

Ribosome is an intercellular structure made of both RNA and protein, and it is the site of protein synthesis in the cell.

4.4. FACTORS FOR SOLUBLE PROTEIN

Numerous soluble proteins cycle through the ribosomes to regulate them, facilitating the process of protein synthesis and increasing the accuracy with which it occurs. In next sections we will examine soluble components' role(s).

4.5. PROTOCOL OF PROTEIN SYNTHESIS

Proteins are synthesized by organisms in all three kingdoms of life utilizing homologous elements and comparable mechanisms. Differences in detail are predicted given the enormous evolutionary range over the past three billion years. In each of the three domains, protein synthesis follows the same four steps (Cai et al., 2015)

- Initiation
- Elongation
- Termination
- Subunit Recycling

Ribosomes are moved with the mRNA when the gene sequence is carried out by conformational changes. The precise linking of tRNAs with their corresponding amino acids & codons in the mRNA that takes place on the ribosome ensures that few errors occur. Proteins called guanosine triphosphatase (GTPase) control the efficiency and progress of each phase (Areta et al., 2013).

4.5.1. Initiation Phase

During initiation, the mRNA's AUG initiator codon is positioned on the ribosome next to the methionine-containing tRNA that serves as the initiator. The two RNAs first form a intricate known as ternary on the smaller ribosomal subunit, called a 70S ribosome in bacteria and an 80S ribosome in eukaryotic cells. Additionally, many eukaryotic initiation factors (eIFs) are soluble proteins that are responsible for coordinating RNA interactions in eukaryotes. Proteins play a smaller role in prokaryotes (abbreviated IF). The following processes occur sequentially in eukaryotes (Steitz & Jakes, 1975):

Step 1. The GTPase eIF-2A and the initiator tRNA Met accumulate into a preinitiation complex on a smaller subunit. (with bound GTP).

Step 2: The second step is for protein initiation components to accumulate on the 5′ cap of mRNA. Using ATP hydrolysis, the RNA helicase eIF-4A removes secondary structure and related proteins from the 5′ ends of the mRNA. These cap recognizing factors cooperate with mRNA's poly (A) tails to produce a cyclical complex that can either stimulate or repress translation initiation. (De Benedetti & Rhoads, 1990).

Step 3: The cap recognition complex guides the mRNA to the pre-initiation complex. Research into the optimal execution of the above first three procedures is ongoing. For instance, the beginning factors and Met-tRNA can't bind to the shorter subunit until after mRNA has already done so.

Step 4: While the P-site initiator tRNA looks for the AUG codon, the smaller subunit and 2 initiation factors open a channel through which the mRNA can move. Hydrolysis of adenosine triphosphate (ATP) is required for this action. However, its precise role is unknown. In bacteria, for example, the localized sequence of the mRNA can affect the specificity, but in eukaryotes, translation normally begins at the first AUG codon. (Stark et al., 2012).

Step 5: Met-tRNA, when it binds to the initiator AUG codon, eIF-2A hydrolyzes the attached GTP.

Step 6: Tiny subunit separates from eIF-2A and other recycling precursors.

Step 7. A big ribosomal subunit binds the tiny ribosomal subunit that contains the Met-tRNA and mRNA. Before polypeptide elongation, a GTPase called eIF-5B hydrolyzes the bound GTP. Phosphorylation of initiation factors is a common step in the first, highly regulated stage of protein combination. For instance, translation initiation factor 2A (eIF-2A) is phosphorylated by cells in response to various stresses. Phosphorylation enhances eIF-2A's binding for eIF-2B, the competing guanine nucleotide exchange factor for the initiator tRNA (Tsokas et al., 2005).

Phosphorylation of eIF-4F develops translation by facilitating the protein's binding to the 5′ caps of messenger RNAs. This approach may compromise the effective translation of specific mRNAs because their 5′ caps have varying degrees of affinity for eIF-4F (Kabat, 1975).

4.5.2. Elongation Phase

Through elongation, the ribosome chooses appropriate aa-tRNAs from an intracellular group based on the codon structure of the translated mRNA. After a peptide bond is formed among the amino group of each subsequent aa-tRNA and the carboxyl group at the C terminus, the ribosome advances to the next codon. Amino acids begin entering into the polypeptide chain according to the codon after the ribosomal subunits are coupled and an initiator tRNA and mRNA are in the correct order. The elongation mechanisms occur in the space among the ribosomal subunits. The mRNA follows a curving path between the subunits as it is threaded through them codon by codon. The aa-tRNAs sequentially bind to three positions between the two ribosomal subunits after entering the gap from one side (Tsokas et al., 2005).

Associations with both subunits allow the tRNA to remain bound to the ribosome as it moves from the A to the P and then to the E site before separation. Base-pairing occurs between the mRNA codons and the tRNAs' anticodons when the latter is bound to the A and P sites, respectively. Peptidyl bonds are formed at the other terminal of the tRNAs, bringing the amino acid on the tRNA to the A site close to the peptidyl chain at the P site of the large subunit. The maturing polypeptide is secreted from the cell over a 10-nm-long channel in the large subunit (Chotiner et al., 2003).

Remember

The physical process of protein folding is what allows a protein chain to assume its physiologically active three-dimensional shape, or "folded" conformation.

Elongation factors, a pair of GTPases, bind around the A site and stimulate subunit movement relative to one another, allowing mRNA and tRNA to travel more freely along the ribosome. A portion of the energy released during GTP hydrolysis is also used to increase accuracy; nonetheless, elongation has become the phase of translation that requires the greatest energy.

The following sentences better describe the four stages of elongation:

1. aa-tRNA binding to ribosome A site
2. Proofreading guarantees proper aa-tRNA binding.
3. Peptide bond formation
4. The peptidyl-tRNA is transferred from the A site of the ribosome to the P site, which moves the mRNA forward by one codon. New structural and spectroscopic investigations of each ribosome will shed more light on the topic.

TIPS

The physical process of protein folding is the transformation of a protein chain into its original 3D structure, usually a "folded" conformation that allows the protein to perform its biological function.

Step 1: Initial process: aa-tRNA binding. The GTPase is activated by a nucleotide exchange factor and loaded with GTP. (eEFX in eukaryotic & EF-Ts in Bacterial cells). This formulates eEF1A to deliver an aa-tRNA to an empty A site on a ribosome. There is sufficient eEF1A-GTP in cells to attach to all of the aa-tRNAs, protecting the sensitive ester link between amino acids.

Step 2: Check for errors. A proofreading process ensures that the precise aa-tRNA remains in the A site while enabling mismatched aa-tRNAs to separate. This "kinetic proofreading mechanism" practices two reactions—GTP hydrolysis related to eEF1A, and the subsequent separation of guanosine diphosphate (GDP)-eEF1A from the aa-tRNA and the ribosome—to identify correct and incorrect aa-tRNAs. If an aa-tRNA anticodon is coupled with the correct mRNA codon, the ribosome triggers GTP hydrolysis, phosphate discharge, a major conformational change, and eEF1A degradation in a matter of milliseconds (Flores-Méndez et al., 2014).

This permits the aminoacyl tail of the aa-tRNA to join with the P site of the large subunit to arrange a peptide bond. aa-rRNAs with feeble, faulty codon-anticodon pairs must exit the A site for eEF1A can hydrolyze GTP and remove tRNA from the aminoacyl.

KEYWORD

Guanosine diphosphate (GDP) is a nucleoside diphosphate. It is an ester of pyrophosphoric acid with the nucleoside guanosine. GDP consists of a pyrophosphate group, a pentose sugar ribose, and the nucleobase guanine.

Step 3: Peptidyl transfer is the third step. The RNA in large subunit is the conserved active site that catalyzes peptide bond development. It involves eliminating water and transferring the carboxyl group from the peptidyl-tRNA's esterified P site to the aa-tRNA's free A site amino group. Peptide bond formation requires precise substrate placement and stability of the conversion state (like enzymes) (Bhadula et al., 2001). The chemical process is like the breakdown of peptide bonds by proteolytic enzymes like chymotrypsin, but in reverse. After developing a new peptide bond, a polypeptide is joined to the tRNA at the A site, with the anticodon arm still linked to the mRNA codon for the small subunit. The antibiotic puromycin can block elongation by functioning as a mimic for either tRNAPhe or tRNATyr. After targeting the esters of carboxyl group of a peptidyl-tRNA in the P site, puromycin halts elongation since there is no good acceptor site for additional peptidyl transference events. It causes the polypeptide chain to be liberated from the ribosome before it is ready. (Gonen et al., 1994).

Step 4: Translocation. The 2nd GTPase elongation factor, or eEF2 in eukaryotes and EF-G in bacteria, stimulates the third connected reaction necessary for a wide-ranging elongation cycle. These GTPases feature a pair of domains analogous to domains 1 and

2 of EF-Tu and three domains that are the size and shape of a tRNA.

Domain 1 fixes GTP and catalyzes its hydrolysis. Domains 3–5 of GTP-eEF2 target an available A site on the ribosome. When attached to a free A site, GTP-eEF2 causes a rotation of the small subunit toward the large subunit of roughly 6 degrees. Reverse rotation of the small subunit, transfer of the peptidyl-tRNA from the A site to the P site, and forward movement of the mRNA by three bases are all stimulated by the dissociation of phosphate and hydrolysis of the GTP bound to eEF2. Forces at this translocation stage are relatively high, at around 13 pN (piconewtons), and are powered by the energy released during peptide bond production. Similarly, as the tRNA is being deacylated in the P site, it is being moved to the exiting (E) site in preparation for its eventual dissociation from the ribosome. Once again, elongation is possible because bound GDP causes eEF2 to dissociate from the A site (Chen et al., 2011).

Additionally, each amino acid with the expanding peptide is forced down an RNA-lined tunnel measuring 10 nm in the large subunit. A 40-residue extended polypeptide, N-terminus first, can squeeze through the opening. The polypeptide undergoes significant folding outside the ribosome, although an α-helix may fit through the tunnel's furthest parts. The large subunit produces peptides that are over 40 residues in length.

To attain an inaccuracy rate of about 1 in 104 improper amino acids, cells must strike a negotiation between accuracy and speed during translation. Since ribosomes can enhance about 20 amino acids to a polypeptide per second at 37°C, they can produce an average-sized protein (300 amino acids) in a matter of 15 seconds. More accurate translation might be possible at a slower rate, but deliberate cellular development might be counterproductive to evolution (Belelovsky et al., 2005).

KEYWORD

Polypeptide synthesis is the biological production of peptides, which are organic compounds in which multiple amino acids are linked via peptide bonds. They are the building blocks of proteins, which are essential to cell function.

4.5.3. Termination Phase

At the 3′-end of the coding arrangement, when the ribosome meets a terminating codon (UAA, UAG, or UGA), termination occurs. Polypeptide synthesis is interrupted when a protein-releasing factor somewhat than an aa-tRNA fixes to the A site on the small subunit of the ribosome. Hence, these releasing factors, RF1 or RF2 in bacteria and eRF1 in eukaryotes, distinguish stop codons and trigger the active ribosome site to break down the peptidyl ester between the polypeptide chain's C-terminal amino acid and the tRNA in the P site. The release of the last polypeptide chain monitors the ribosome. Using the energy from GTP hydrolysis, a GTPase recycles the ribosomal subunits and mRNA after they split apart and begins to translate new mRNAs (Yabuki et al., 2007).

4.6. SPONTANOUS PROTEIN FOLDING

An entirely new protein only starts to form once the translation process has ended. A polypeptide begins interacting with its new surroundings as it is formed. Its N-terminus exits the large ribosomal subunit tunnel when it is about 40 residues long and reaches the cytoplasm, where it has to fold into a three-dimensional structure and find its appropriate cellular position. (Anfinsen & Scheraga, 1975).

KEYWORD

Lipid bilayer is a thin polar membrane made of two layers of lipid molecules. These membranes are flat sheets that form a continuous barrier around all cells.

Because the form of gathered proteins and how they are folded are determined by the amino acid arrangement, folding can happen spontaneously underneath the right conditions. These situations are aqueous solvents at a particular temperature, a pH of 7, and mild ionic strength for the soluble proteins. Transmembrane proteins fold considerably differently in a lipid bilayer. High temperatures, pH extremes, or large amounts of urea or guanidine can deactivate small soluble proteins in test tube tests. Proteins that have been denatured are groupings of polymers that no longer possess secondary structures (Baker, 2000).

It takes milliseconds to microseconds for small, denatured polypeptides to fold back into their instinctive three-dimensional structures after being subjected to physiological conditions. (Proteins that need the isomerization of prolines, like collagen, fold significantly quite gradually.) A polypeptide converges toward a single low-energy native state from multiple initial denatured states after being directed by energy from various noncovalent bonds and the hydrophobic effect. (Finkelstein & Galzitskaya, 2004).

Proteins would never fold if each possible path to the native state were sampled independently since numerous potential methods exist. Both theory and experiment point to several mechanisms that are involved in folding, including the emergence of a variety of dispersed states of transition with secondary structural elements, certain twists, and hydrophobic bonds present in the native protein's central region. Nevertheless, since the free energy terrian for folding contains hills and valleys, proteins can become confined in partially folded shapes. Many proteins crease *in vivo* throughout biosynthesis naturally and without assistance. Compacting begins when the N-terminus of the developing polypeptide exits the ribosome. There are advantages and disadvantages to the extremely slow "co-translational folding" vectorial structure. Early folding has the advantage of limiting the routes to the pleated state, which may help to explain why some proteins coil more efficiently in biosynthesis from the folded state than from the inactive form. However, vectorial folding precludes connections among proteins' N-terminal and C-terminal areas until they have left the ribosome. In folded proteins, these interactions are common. (Rose et al., 2006).

KEYWORD

Biosynthesis is the generation of natural products through enzymatic reactions, as in cellular metabolism. Successive enzymatic reactions by a number of enzymes are generally required to achieve a single biologically active compound.

Massive proteins are very challenging to fold, especially in the cytoplasm's dense environment, where improperly folded proteins might uncover hydrophobic regions normally tucked away in the native protein's center. These exposed center components may be combined permanently before folding is complete. Many recently manufactured native proteins require support to avoid permanent deactivation, aggregation, or breakdown by proteolysis during folding.

Often, misfolded mutant proteins are to blame for human illness. For instance, the utmost reason for cystic fibrosis is a genetic mutation that alters the codon for just one amino acid in the cystic fibrosis transmembrane regulator (CFTR), so preventing the protein from folding properly. Misfolded proteins in blistering skin diseases, hypertrophic cardiomyopathies, and other "dominant negative" circumstances not only have no function but also contaminate the synthesis of natural proteins. Proteins folding into abnormal shapes is what causes amyloid and prion diseases (Luengo et al., 2019).

4.7. PROTIEN FOLDING WITH THE ASSISTANCE OF CHAPERONE

Proteins are complex biomolecules that play a crucial role in almost every biological process. To function properly, a protein must be in its correct three-dimensional shape, which is determined by its amino acid sequence. The process of achieving this final shape is called protein folding.

Protein folding is the process by which a protein assumes its functional, three-dimensional shape through the formation of specific, non-covalent interactions between its amino acid residues.

The correct folding of proteins is essential for their proper function. Misfolded or unfolded proteins can lead to a wide range of diseases, including Alzheimer's, Huntington's, and Parkinson's diseases. Protein folding is a complex and error-prone process, and cells have evolved various mechanisms to assist in the folding of proteins. One such mechanism is the use of chaperones, which are a group of specialized proteins that facilitate the folding of other proteins. Chaperones act by binding to unfolded or partially folded proteins, preventing them from aggregating or being degraded, and assisting them in reaching their correct, functional conformation. Chaperones are particularly important in conditions of cellular stress, such as high temperatures, where protein folding is impaired and proteins are more prone to misfolding or aggregation. Chaperones are present in all living organisms, and different types of chaperones have been identified, including molecular chaperones, chaperonins, and co-chaperones. These chaperones work together in a coordinated manner to assist in protein folding and prevent the accumulation of misfolded or unfolded proteins.

KEYWORD

Protein folding is a process by which a polypeptide chain folds to become a biologically active protein in its native 3D structure.

In summary, protein folding is an essential process that ensures the proper function of proteins. Chaperones play a critical role in protein folding by assisting proteins in achieving their functional conformation, and preventing the accumulation of misfolded or unfolded proteins. Understanding the mechanisms and regulation of protein folding with chaperones is a crucial area of research in biochemistry and has implications for the development of therapeutics for protein folding-related diseases.

Protein folding is a complex process that requires the assistance of molecular chaperones to ensure proper folding and prevent aggregation of unfolded or misfolded proteins. The mechanisms of protein folding with chaperones can be broadly categorized into two types: co-translational folding and post-translational folding. Co-translational folding occurs during the synthesis of the protein on ribosomes. The nascent polypeptide chain begins to fold as it emerges from the ribosome and is co-translationally bound by chaperones, which assist in the folding process. The chaperones help the protein to fold into its native conformation by preventing misfolding, aggregation, or degradation. The ribosome-associated chaperones include the trigger factor (TF), Hsp70, and Hsp40. Trigger factor is the first chaperone that binds the nascent chain as it emerges from the ribosome and helps to stabilize the structure of the newly synthesized protein. Hsp70 and Hsp40 chaperones bind to the nascent chain after trigger factor and assist in the folding process by preventing aggregation of misfolded intermediates.

Examples of co-translational folding chaperones include DnaK, GroEL, and GroES. Post-translational folding occurs after the completion of protein synthesis. Once the protein is released from the ribosome, it is subjected to the cellular environment, which can be harsh and prone to misfolding. Chaperones help to prevent misfolding and aggregation by binding to the protein and guiding it towards its native conformation. The post-translational chaperones include Hsp60, Hsp70, and Hsp90. Hsp60 chaperonins form a large chamber-like structure where the protein can fold without interference from the surrounding environment. Hsp70 and Hsp90 chaperones bind to the protein and assist in the folding process by stabilizing intermediates and preventing aggregation.

ACTIVITY 4.1.

Differentiate between Hsp 70 and Hsp 90 chaperones with a chart.

The roles of chaperones in co-translational folding and post-translational folding differ in terms of the stage of the folding process where they are involved. In co-translational folding, chaperones bind to the nascent polypeptide chain as it emerges from the ribosome and assist in the folding process by stabilizing intermediates and preventing aggregation. In post-translational folding, chaperones bind to the protein after it has been synthesized and help to guide it towards its native conformation by preventing misfolding and aggregation. There are numerous examples of chaperones involved in protein folding. For instance, the GroEL and GroES chaperones are involved in co-translational folding of bacterial proteins. These chaperones work together to form a large, barrel-shaped structure that accommodates the folding protein and prevents aggregation. Hsp70 chaperones, such as DnaK in bacteria and BiP in eukaryotes, are involved in both co-translational and post-translational folding of proteins. They bind to the protein and assist in the folding process by preventing aggregation and stabilizing intermediates. Hsp90 chaperones are also involved in post-translational folding and are important for the folding of many signal transduction proteins, including steroid hormone receptors and protein kinases.

SUMMARY

- Mechanism for Protein Synthesis

  According to this chapter's summary, this structure is made up of ribosomes, amino acids, mRNA, tRNA, and aminoacyl tRNA synthetases. The genetic code is contained in mRNA, which serves as a template.

- Spontaneous Protein folding

  For protein folding to be regarded as a spontaneous reaction, it must occur within a cell under thermodynamically favorable conditions. Since it is well known that the spontaneous response of protein folding is an example, we may infer that this method has to have a negative Gibbs free energy value. Protein folding requires the Gibbs free energy, equal to both enthalpy and entropy.

- Chaperone-Assisted Protein Folding

  Adding to and maintaining either entirely or partially unfolded protein polypeptides, chaperones inhibit the aggregation and incorrect folding of protein chains. An intact polypeptide chain is synthesized at the end of this step. They also prevent unfolding polypeptide chains from folding unsteady while they are transported into the subcellular organelles.

REVIEW QUESTIONS

1. What do you understand by protein synthesis and folding?
2. Explain the transfer RNA in detail.
3. Explain the protocol of protein synthesis along with its three phases.
4. What is spontaneous protein folding?
5. Differentiate between Hsp70 chaperones and Hsp90 chaperones.

MULTIPLE CHOICE QUESTIONS

1. Which amino acids are buried deep inside a protein structure?

a. Either hydrophobic or hydrophilic
b. Both hydrophobic and hydrophilic
c. Hydrophilic
d. Hydrophobic

2. Which of the following amino acids are likely to be present on the surface of a globular protein?

a. Both hydrophobic and hydrophilic
b. Either hydrophobic or hydrophilic

c. Hydrophobic

d. Hydrophilic

3. Which of the following class of proteins helps a newly made protein molecule to fold properly?

a. Mitochondrial proteins

b. Ribosomal proteins

c. Immunoglobulins

d. Chaperones

4. Which of the following is the correct sequence of events in protein synthesis?

a. Transcription, translation, post-translational modifications

b. Translation, transcription, post-translational modifications

c. Transcription, post-translational modifications, translation

d. Translation, post-translational modifications, transcription

5. The process of transcription involves:

a. The synthesis of RNA from DNA

b. The synthesis of DNA from RNA

c. The folding of proteins into their functional form

d. The degradation of proteins into amino acids

Answers to Multiple Choice Questions

1. (d) 2. (d) 3. (d) 4. (a) 5. (a)

REFERENCES

1. Anfinsen, C. B., & Scheraga, H. A. (1975). Experimental and theoretical aspects of protein folding. *Advances in Protein Chemistry*, *29*, 205-300.
2. Areta, J. L., Burke, L. M., Ross, M. L., Camera, D. M., West, D. W., Broad, E. M., ... & Coffey, V. G. (2013). Timing and distribution of protein ingestion during prolonged recovery from resistance exercise alters myofibrillar protein synthesis. *The Journal of Physiology*, *591*(9), 2319-2331.
3. Baker, D. (2000). A surprising simplicity to protein folding. *Nature*, *405*(6782), 39-42.
4. Belelovsky, K., Elkobi, A., Kaphzan, H., Nairn, A. C., & Rosenblum, K. (2005). A molecular switch for translational control in taste memory consolidation. *European Journal of Neuroscience*, *22*(10), 2560-2568.
5. Bhadula, S. K., Elthon, T. E., Habben, J. E., Helentjaris, T. G., Jiao, S., & Ristic, Z. (2001). Heat-stress induced synthesis of chloroplast protein synthesis elongation factor (EF-Tu) in a heat-tolerant maize line. *Planta*, *212*(3), 359-366.

6. Cai, Q., Hanson, J. A., Steiner, A. R., Tran, C., Masikat, M. R., Chen, R., ... & Yin, G. (2015). A simplified and robust protocol for immunoglobulin expression in E scherichia coli cell-free protein synthesis systems. *Biotechnology Progress*, *31*(3), 823-831.
7. Chen, C. Y., Fang, H. Y., Chiou, S. H., Yi, S. E., Huang, C. Y., Chiang, S. F., ... & Chow, K. C. (2011). Sumoylation of eukaryotic elongation factor 2 is vital for protein stability and anti-apoptotic activity in lung adenocarcinoma cells. *Cancer science*, *102*(8), 1582-1589.
8. Chotiner, J. K., Khorasani, H., Nairn, A. C., O'dell, T. J., & Watson, J. B. (2003). Adenylyl cyclase-dependent form of chemical long-term potentiation triggers translational regulation at the elongation step. *Neuroscience*, *116*(3), 743-752.
9. Das, D., Das, A., Samanta, D., Ghosh, J., Dasgupta, S., Bhattacharya, A., ... & Das Gupta, C. (2008). Role of the ribosome in protein folding. *Biotechnology Journal: Healthcare Nutrition Technology*, *3*(8), 999-1009.
10. De Benedetti, A., & Rhoads, R. E. (1990). Overexpression of eukaryotic protein synthesis initiation factor 4E in HeLa cells results in aberrant growth and morphology. *Proceedings of the National Academy of Sciences*, *87*(21), 8212-8216.
11. Finkelstein, A. V., & Galzitskaya, O. V. (2004). Physics of protein folding. *Physics of Life Reviews*, *1*(1), 23-56.
12. Flores-Méndez, M., Ramírez, D., Alamillo, N., Hernández-Kelly, L. C., Del Razo, L. M., & Ortega, A. (2014). Fluoride exposure regulates the elongation phase of protein synthesis in cultured Bergmann glia cells. *Toxicology Letters*, *229*(1), 126-133.
13. Garcia-Esparcia, P., Hernández-Ortega, K., Koneti, A., Gil, L., Delgado-Morales, R., Castaño, E., ... & Ferrer, I. (2015). Altered machinery of protein synthesis is region- and stage-dependent and is associated with α-synuclein oligomers in Parkinson's disease. *Acta neuropathologica communications*, *3*(1), 1-25.
14. Gast, V., Campbell, K., Picazo, C., Engqvist, M., Siewers, V., & Molin, M. (2021). The Yeast eIF2 Kinase Gcn2 Facilitates H2O2-Mediated Feedback Inhibition of Both Protein Synthesis and Endoplasmic Reticulum Oxidative Folding during Recombinant Protein Production. *Applied and Environmental Microbiology*, *87*(15), 1-21.
15. Gonen, H., Smith, C. E., Siegel, N. R., Kahana, C., Merrick, W. C., Chakraburtty, K., ... & Ciechanover, A. (1994). Protein synthesis elongation factor EF-1 alpha is essential for ubiquitin-dependent degradation of certain N alpha-acetylated proteins and may be substituted for by the bacterial elongation factor EF-Tu. *Proceedings of the National Academy of Sciences*, *91*(16), 7648-7652.
16. Groeneveld, H. E. R. M. A. N., Thimon, K. A. F. I. I. N. E., & Van Duin, J. (1995). Translational control of maturation-protein synthesis in phage MS2: a role for the kinetics of RNA folding?. *RNA*, *1*(1), 79.
17. Horowitz, S., Salmon, L., Koldewey, P., Ahlstrom, L. S., Martin, R., Quan, S., ... & Bardwell, J. C. (2016). Visualizing chaperone-assisted protein folding. *Nature Structural & Molecular Biology*, *23*(7), 691-697.

18. Houry, W. A. (2001). Chaperone-assisted protein folding in the cell cytoplasm. *Current Protein and Peptide Science*, *2*(3), 227-244.
19. Johnson, E. C., Durek, T., & Kent, S. B. (2006). Total chemical synthesis, folding, and assay of a small protein on a water-compatible solid support. *Angewandte Chemie*, *118*(20), 3361-3365.
20. Kabat, D. (1975). Potentiation of hemoglobin messenger ribonucleic acid. A step in protein synthesis initiation involving interaction of messenger with 18 S ribosomal ribonucleic acid. *Journal of Biological Chemistry*, *250*(15), 6085-6092.
21. Kowalak, J. A., Dalluge, J. J., McCloskey, J. A., & Stetter, K. O. (1994). The role of posttranscriptional modification in stabilization of transfer RNA from hyperthermophiles. *Biochemistry*, *33*(25), 7869-7876.
22. Kramer, G., Boehringer, D., Ban, N., & Bukau, B. (2009). The ribosome as a platform for co-translational processing, folding and targeting of newly synthesized proteins. *Nature Structural & Molecular Biology*, *16*(6), 589-597.
23. Laufen, T., Mayer, M. P., Beisel, C., Klostermeier, D., Mogk, A., Reinstein, J., & Bukau, B. (1999). Mechanism of regulation of hsp70 chaperones by DnaJ cochaperones. *Proceedings of the National Academy of Sciences*, *96*(10), 5452-5457.
24. Li, B., Lai, T., Qin, G., & Tian, S. (2010). Ambient pH stress inhibits spore germination of *Penicillium expansum* by impairing protein synthesis and folding: a proteomic-based study. *Journal of Proteome Research*, *9*(1), 298-307.
25. Liu, S., Wang, X., Sun, F., Zhang, J., Feng, J., Liu, H., ... & Liu, Z. (2013). RNA-Seq reveals expression signatures of genes involved in oxygen transport, protein synthesis, folding, and degradation in response to heat stress in catfish. *Physiological Genomics*, *45*(12), 462-476.
26. Lorimer, G. H. (1996). A quantitative assessment of the role of the chaperonin proteins in protein folding *in vivo*. *The FASEB journal*, *10*(1), 5-9.
27. Lowe, T. M., & Chan, P. P. (2016). tRNAscan-SE On-line: integrating search and context for analysis of transfer RNA genes. *Nucleic Acids Research*, *44*(1), 54-57.
28. Lowe, T. M., & Eddy, S. R. (1997). tRNAscan-SE: a program for improved detection of transfer RNA genes in genomic sequence. *Nucleic Acids Research*, *25*(5), 955-964.
29. Luengo, T. M., Mayer, M. P., & Rüdiger, S. G. (2019). The Hsp70–Hsp90 chaperone cascade in protein folding. *Trends in Cell Biology*, *29*(2), 164-177.
30. Mayer, M. P. (2005). Recruitment of Hsp70 chaperones: a crucial part of viral survival strategies. *Reviews of Physiology, Biochemistry and Pharmacology*, 1, 1-46.
31. O'donnell, C. W., & Lis, M. (2006). The Trigger factor chaperone. *People Csail Mit Edu*, 1, 1-21.
32. Pawlowska, Z., Baranska, P., Jerczynska, H., Koziolkiewicz, W., & Cierniewski, C. S. (2005). Heat shock proteins and other components of cellular machinery for protein synthesis are up-regulated in vascular endothelial cell growth factor-activated human endothelial cells. *Proteomics*, *5*(5), 1217-1227.

33. Rose, G. D., Fleming, P. J., Banavar, J. R., & Maritan, A. (2006). A backbone-based theory of protein folding. *Proceedings of the National Academy of Sciences*, *103*(45), 16623-16633.

34. Sánchez, E. G., Quintas, A., Nogal, M., Castelló, A., & Revilla, Y. (2013). African swine fever virus controls the host transcription and cellular machinery of protein synthesis. *Virus Research*, *173*(1), 58-75.

35. Schopf, F. H., Biebl, M. M., & Buchner, J. (2017). The HSP90 chaperone machinery. *Nature Reviews Molecular cell biology*, *18*(6), 345-360.

36. Skach, W. R. (2009). Cellular mechanisms of membrane protein folding. *Nature Structural & Molecular Biology*, *16*(6), 606-612.

37. Stark, M., Lukaszuk, J., Prawitz, A., & Salacinski, A. (2012). Protein timing and its effects on muscular hypertrophy and strength in individuals engaged in weight-training. *Journal of the International Society of Sports Nutrition*, *9*(1), 1-8.

38. Steitz, J. A., & Jakes, K. (1975). How ribosomes select initiator regions in mRNA: base pair formation between the 3'terminus of 16S rRNA and the mRNA during initiation of protein synthesis in *Escherichia coli*. *Proceedings of the National Academy of Sciences*, *72*(12), 4734-4738.

39. Takei, N., Inamura, N., Kawamura, M., Namba, H., Hara, K., Yonezawa, K., & Nawa, H. (2004). Brain-derived neurotrophic factor induces mammalian target of rapamycin-dependent local activation of translation machinery and protein synthesis in neuronal dendrites. *Journal of Neuroscience*, *24*(44), 9760-9769.

40. Tsokas, P., Grace, E. A., Chan, P., Ma, T., Sealfon, S. C., Iyengar, R., & Blitzer, R. D. (2005). Local protein synthesis mediates a rapid increase in dendritic elongation factor 1A after induction of late long-term potentiation. *Journal of Neuroscience*, *25*(24), 5833-5843.

41. Waudby, C. A., Dobson, C. M., & Christodoulou, J. (2019). Nature and regulation of protein folding on the ribosome. *Trends in Biochemical Sciences*, *44*(11), 914-926.

42. Woodson, S. A. (2011). RNA folding pathways and the self-assembly of ribosomes. *Accounts of Chemical Research*, *44*(12), 1312-1319.

43. Woolhead, C. A., McCormick, P. J., & Johnson, A. E. (2004). Nascent membrane and secretory proteins differ in FRET-detected folding far inside the ribosome and in their exposure to ribosomal proteins. *Cell*, *116*(5), 725-736.

44. Yabuki, T., Motoda, Y., Hanada, K., Nunokawa, E., Saito, M., Seki, E., & Yokoyama, S. (2007). A robust two-step PCR method of template DNA production for high-throughput cell-free protein synthesis. *Journal of Structural and Functional Genomics*, *8*(4), 173-191.

45. Yewdell, J. W. (2001). Not such a dismal science: the economics of protein synthesis, folding, degradation and antigen processing. *Trends in Cell Biology*, *11*(7), 294-297.

46. Zhang, G., Hubalewska, M., & Ignatova, Z. (2009). Transient ribosomal attenuation coordinates protein synthesis and co-translational folding. *Nature Structural & Molecular Biology*, *16*(3), 274-280.

47. Zuiderweg, E. R., Hightower, L. E., & Gestwicki, J. E. (2017). The remarkable multivalency of the Hsp70 chaperones. *Cell Stress and Chaperones*, *22*(2), 173-189.

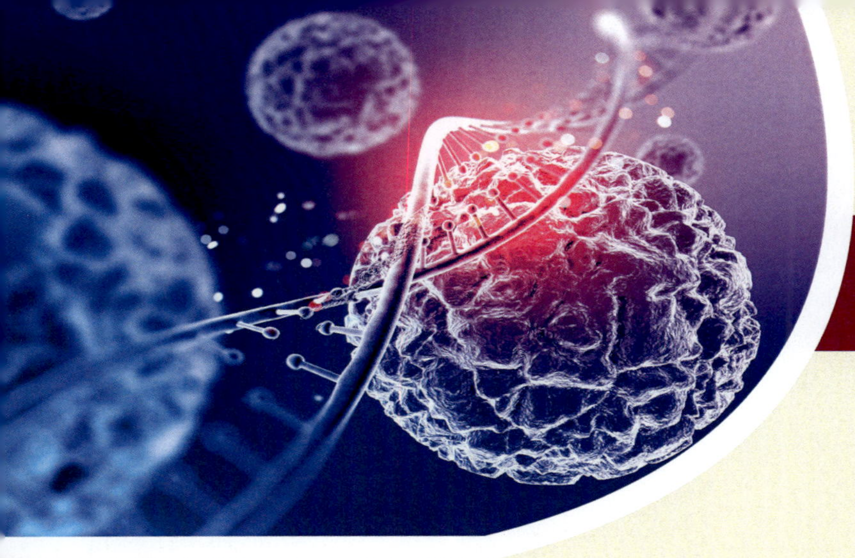

CHAPTER 5

GENE EXPRESSION

UNIT INTRODUCTION

Gene expression is a fundamental process that plays a crucial role in the functioning of all living organisms. It is the process by which the information stored in a gene is used to create functional gene products such as proteins. This chapter aims to provide an understanding of gene expression, its definition, importance in cellular processes, and an overview of the key molecules involved, including DNA, RNA, and protein synthesis. Gene expression refers to the process by which the information encoded in a gene is utilized to produce a functional gene product. This process involves the conversion of the genetic information stored in DNA into a functional molecule, usually a protein or RNA molecule, that performs specific tasks within the cell. Gene expression is a complex, tightly regulated process that ensures that the right genes are expressed at the right time and in the right amounts. Gene expression is essential for the proper functioning of cells and the development of multicellular organisms. It controls various cellular processes, including growth, development, differentiation, and response to environmental stimuli. Gene expression allows cells to adapt to changing conditions and maintain homeostasis (Brazma & Vilo, 2000). The regulation of gene expression is crucial for cell specialization and the formation of different tissues and organs during development (Wingender et al., 2001). For example, the expression of specific genes in a developing embryo determines the formation of various body structures, such as limbs, organs, and the nervous system. In addition to development, gene expression is also vital for the maintenance of normal cellular functions. Genes involved in metabolism, DNA repair, cell cycle control, and signal transduction pathways are all regulated to ensure proper cellular processes. Any disruption in gene expression can lead to diseases, including cancer, where genes involved in cell growth and division are improperly regulated (Audic & Claverie, 1997).

Learning Objectives

At the end of this chapter, readers will be able to:

- Define gene expression and understand its fundamental role in the synthesis of proteins and cellular processes.
- Explain the processes of transcription and translation, including the structure and function of RNA polymerase, ribosomes, and the genetic code.
- Describe the various steps involved in transcription, including initiation, elongation, and termination, and understand the importance of post-transcriptional modifications in RNA processing.
- Discuss the significance of post-translational modifications in proteins and their role in regulating protein function, stability, and localization.
- Identify and explain the different mechanisms of gene regulation, including transcriptional regulation through gene regulatory elements, such as enhancers, promoters, and silencers, as well as the involvement of transcription factors.
- Understand the impact of chromatin structure and epigenetic modifications on gene expression, and their role in cell differentiation and development.
- Explore the role of non-coding RNAs, including microRNAs, in gene regulation and their potential as therapeutic targets.
- Describe the techniques used in gene expression analysis, such as Northern blotting, RT-PCR, microarray analysis, RNA sequencing, and gene expression profiling.
- Understand the importance of bioinformatics tools in analyzing and interpreting gene expression data, including identifying differentially expressed genes, functional enrichment analysis, and pathway analysis.
- Recognize the significance of gene expression in development and disease, including the regulation of gene expression during embryonic development and the changes in gene expression observed in cancer and other diseases.

Key Terms

- Gene Expression
- Transcription
- Translation
- RNA Polymerase
- Ribosomes
- Genetic Code
- Post-Transcriptional Modifications
- Gene Regulatory Elements
- Transcription Factors
- Chromatin Structure
- Epigenetic Modifications
- Gene Therapy

5.1. OVERVIEW OF DNA, RNA, AND PROTEIN SYNTHESIS

To understand gene expression, it is essential to grasp the basic concepts of DNA, RNA, and protein synthesis. DNA, or deoxyribonucleic acid, is the genetic material that carries the instructions for the development, functioning, and reproduction of all living organisms. It consists of a double helix structure made up of nucleotide building blocks. DNA serves as a template for the synthesis of RNA, a single-stranded nucleic acid molecule. RNA plays various roles in gene expression, including the transfer of genetic information from DNA to the site of protein synthesis. The three main types of RNA involved in gene expression are messenger RNA (mRNA), ribosomal RNA (rRNA), and transfer RNA (tRNA) (Schneider, 2012).

Remember

Gene expression is a tightly regulated process that involves the conversion of genetic information from DNA to RNA and eventually to proteins. It is crucial for various cellular processes and plays a significant role in development, disease, and therapeutic interventions.

Protein synthesis, also known as translation, is the process by which the genetic information encoded in mRNA is used to synthesize proteins. This process occurs in cellular structures called ribosomes, which consist of rRNA and proteins. During translation, tRNA molecules bind to specific amino acids and deliver them to the ribosome, where they are joined together to form a protein chain based on the mRNA sequence. In summary, gene expression is a fundamental process that underlies the functioning of living organisms. It involves the conversion of genetic information stored in DNA into functional gene products, such as proteins or RNA molecules. Gene expression is crucial for cellular processes, including development, growth, and response to environmental cues. Understanding the basics of DNA, RNA, and protein synthesis is essential for comprehending the intricacies of gene expression (Fishburn et al., 2005).

5.2. TRANSCRIPTION: FROM DNA TO RNA

Transcription is a critical process in gene expression that involves the synthesis of RNA molecules using DNA as a template. This section will explore the different stages of transcription, including the structure and function of RNA polymerase, initiation, elongation, termination, and the subsequent post-transcriptional modifications of RNA (Coulon et al., 2104).

5.2.1. Structure and Function of RNA Polymerase

RNA polymerase is the enzyme responsible for catalyzing the synthesis of RNA from a DNA template during transcription. In bacteria, the primary RNA polymerase is composed of multiple subunits, including a core enzyme and a sigma factor, which helps in the recognition and binding of specific DNA sequences known as promoters. The core enzyme consists of several subunits, including two α subunits, one β subunit, one β' subunit, and one ω subunit. These subunits work together to catalyze the formation of phosphodiester bonds between nucleotides during RNA synthesis. The sigma factor, on the other hand, is involved in the recognition of the promoter sequence and initiation of transcription. In eukaryotes, there are multiple types of RNA polymerases responsible for transcribing different classes of RNA. RNA polymerase II is the key enzyme involved in transcribing protein-coding genes. It consists of multiple subunits, including the largest subunit known as RPB1. Additionally, several other proteins called transcription factors are required for the proper functioning of RNA polymerase II (Wandzik et al., 2020).

The overall function of RNA polymerase is to bind to a specific DNA sequence called the promoter, unwind the DNA helix, and initiate the synthesis of RNA by adding nucleotides complementary to the DNA template strand (Johnson et al., 1987).

KEYWORD

Gene expression is the process by which the information encoded in a gene is turned into a function. This mostly occurs via the transcription of RNA molecules that code for proteins or non-coding RNA molecules that serve other functions.

5.2.2. Initiation of Transcription

Transcription initiation is a highly regulated process that begins with the binding of RNA polymerase to the promoter region of the DNA. In bacteria, the sigma factor helps in the recognition of the promoter sequence, and once the RNA polymerase-sigma complex binds to the DNA, the DNA strands are separated to form a transcription bubble. In eukaryotes, transcription initiation is more complex. Specific transcription factors, including the general transcription factor TFIIA, TFIIB, TFIID, TFIIE, TFIIF, and TFIIH, bind to the promoter region and help in the assembly of RNA polymerase II at the transcription start site. The general transcription factor TFIID recognizes the TATA box sequence in the promoter and recruits other factors to form the pre-initiation complex (Thiffault et al., 2015).

Once the RNA polymerase is positioned at the transcription start site, it undergoes a series of conformational changes, and the

DNA is unwound to expose the template strand for RNA synthesis.

Elongation and Termination of Transcription

After transcription initiation, the RNA polymerase moves along the DNA template, synthesizing an RNA molecule in a process known as elongation. As the RNA polymerase advances, the DNA helix reforms behind it, and the newly synthesized RNA molecule is released from the enzyme. During elongation, the RNA polymerase maintains a stable interaction with the DNA template and synthesizes an RNA molecule that is complementary to the coding strand of the DNA. The RNA molecule is built in the 5' to 3' direction, adding nucleotides one by one to the growing RNA chain (Barman et al., 2022).

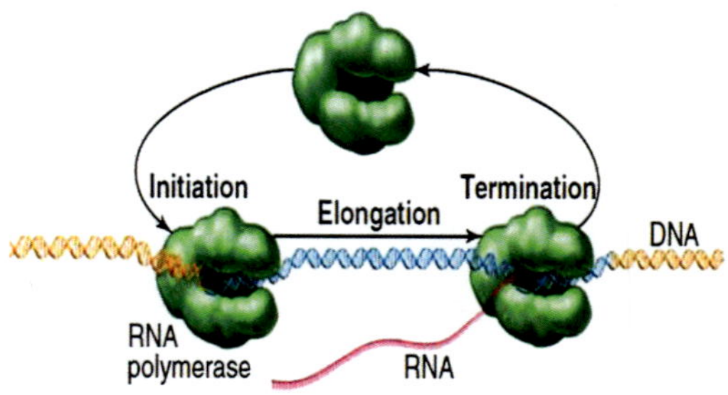

Figure 5.1. The figure illustrates the transcription cycle (Source: Basic medical key, Creative Commons License).

Termination of transcription occurs once the RNA polymerase reaches a specific termination signal on the DNA template. In bacteria, there are two types of termination mechanisms: intrinsic termination and rho-dependent termination. Intrinsic termination involves the formation of a stem-loop structure in the RNA transcript, followed by a string of uracil residues. This structure causes the RNA polymerase to pause and eventually dissociate from the DNA template (Barman et al., 2022).

In rho-dependent termination, a protein called rho factor binds to the growing RNA transcript and moves along it, using ATP energy. When the rho factor reaches the RNA polymerase, it causes the enzyme to detach from the DNA template, resulting in termination of transcription. In eukaryotes, termination of transcription is a more complex process. Two main mechanisms are involved: polyadenylation and termination by RNA polymerase II pausing. Polyadenylation involves the addition of a poly(A) tail to the 3' end of the RNA molecule. This process is mediated by specific proteins that recognize a consensus sequence in the RNA transcript and cleave it downstream, followed by the addition of a string of adenine nucleotides (Forget et al., 2010).

Termination by RNA polymerase II pausing occurs when the RNA polymerase II complex pauses at a termination site. This pause can be influenced by specific DNA sequences and transcriptional regulatory factors. The pausing of RNA polymerase II leads to the release of the RNA transcript and dissociation of the enzyme from the DNA template.

5.2.3. Post-transcriptional Modifications of RNA

After transcription, the newly synthesized RNA molecule, known as the primary transcript or pre-mRNA in eukaryotes, undergoes various post-transcriptional modifications to become a mature, functional RNA molecule. These modifications include:

- RNA Splicing: In eukaryotes, many genes contain non-coding regions called introns within the coding regions called exons. RNA splicing is the process of removing introns and joining the exons to produce a continuous, protein-coding mRNA. This process is carried out by a complex called the spliceosome, composed of small nuclear ribonucleoproteins (snRNPs) and other proteins. Alternative splicing can lead to the generation of different mRNA isoforms from a single gene, increasing the diversity of proteins that can be produced.
- Addition of a 5' Cap: The 5' end of the primary transcript is modified by the addition of a modified nucleotide cap. This cap, known as the 7-methylguanosine cap, provides stability to the mRNA molecule, facilitates its transport out of the nucleus, and plays a role in translation initiation.
- Addition of a 3' Poly(A) Tail: The 3' end of the primary transcript is modified by the addition of a string of adenine nucleotides called the poly(A) tail. This modification protects the mRNA from degradation and is involved in the export of the mRNA from the nucleus and translation initiation.
- RNA Editing: In some cases, the sequence of the RNA molecule can be modified after transcription through a process called RNA editing. This includes the insertion, deletion, or modification of specific nucleotides, altering the information encoded by the DNA. RNA editing is observed in various organisms and can have significant impacts on gene expression and protein function.

These post-transcriptional modifications are essential for the stability, processing, and functionality of RNA molecules. They

Did you Know?

RNA polymerase can transcribe DNA at an astonishing speed of about 40 nucleotides per second! This rapid pace allows for efficient and timely production of RNA molecules during gene expression.

enable the precise control of gene expression and the generation of diverse RNA and protein isoforms with different functions (Lane & Darst, 2010). Transcription is a fundamental process in gene expression that involves the synthesis of RNA molecules from a DNA template. RNA polymerase, the key enzyme responsible for transcription, undergoes initiation, elongation, and termination stages. Post-transcriptional modifications, including RNA splicing, addition of a 5' cap and 3' poly(A) tail, and RNA editing, further process the RNA molecule to produce a mature, functional RNA species. These processes are crucial for the regulation and diversification of gene expression, enabling the cell to produce a wide array of proteins and functional RNAs to carry out its biological functions (White, 2005).

5.3. RNA PROCESSING AND MATURATION

RNA processing and maturation are crucial steps in gene expression that occur after transcription and involve various modifications and quality control mechanisms. This section will delve into the processes of RNA splicing and alternative splicing, the addition of a 5' cap and 3' poly(A) tail, RNA editing and modifications, and the quality control mechanisms involved in RNA processing (Venkataraman et al., 2018).

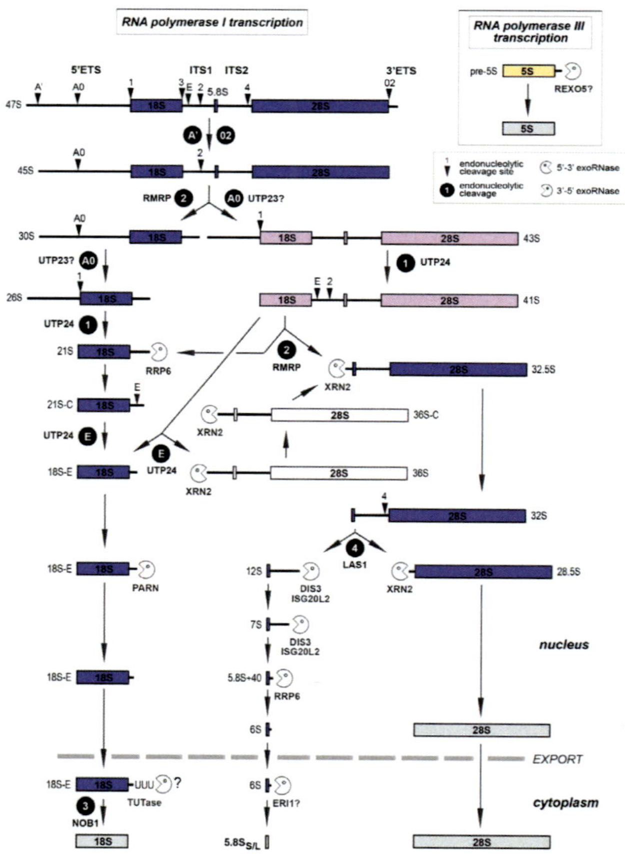

Figure 5.2. Ribosomal RNA production (Source: Aubert et al., Creative Commons License).

5.3.1. RNA Splicing and Alternative Splicing

RNA splicing is a process in which non-coding regions called introns are removed from the primary transcript (pre-mRNA) and the remaining coding regions, known as exons, are joined together to produce a mature mRNA molecule. This process occurs in the nucleus of eukaryotic cells and is carried out by a complex called the spliceosome (Koonin, 1991). The spliceosome is composed of small nuclear ribonucleoproteins (snRNPs) and other proteins. These components recognize specific sequences at the exon-intron boundaries, called splice sites, to accurately identify the introns to be removed. The spliceosome catalyzes two transesterification reactions, known as splicing reactions, to excise the intron and ligate the exons together. Alternative splicing is a phenomenon in which different combinations of exons within a pre-mRNA can be joined together, leading to the production of multiple mRNA isoforms from a single gene. This process allows for increased protein diversity by generating transcripts with different combinations of coding regions. Alternative splicing is regulated by various factors, including splicing enhancers and repressors, as well as the spliceosome itself.

KEYWORD

Spliceosome is a large RNA-protein complex that catalyses the removal of introns from nuclear pre-mRNA.

5.3.2. Addition of 5' Cap and 3' Poly(A) Tail

The primary transcript undergoes modifications at both ends to facilitate its stability, transport, and translation. At the 5' end, a modified nucleotide cap, known as the 5' cap, is added. This cap structure consists of a 7-methylguanosine residue linked to the mRNA via a unique 5' to 5' triphosphate linkage. The 5' cap plays essential roles in protecting the mRNA from degradation, enhancing mRNA stability, and promoting translation initiation (Iyer et al., 2003).

At the 3' end, a string of adenine nucleotides called the poly(A) tail is added. This process, known as polyadenylation, involves the cleavage of the pre-mRNA downstream of a specific signal sequence, followed by the addition of approximately 200 adenine residues. The poly(A) tail contributes to mRNA stability, facilitates mRNA export from the nucleus, and plays a role in translation and mRNA decay. Both the 5' cap and the 3' poly(A) tail are involved in the recognition and processing of the mRNA by cellular machinery and are important for the efficient and regulated expression of genes (Lisica et al., 2016).

5.3.3. RNA Editing and Modifications

In addition to splicing and capping, RNA molecules can undergo various modifications, including editing and chemical modifications of individual nucleotides. RNA editing is a process that alters the nucleotide sequence of RNA molecules, resulting in changes in the encoded information. This process can involve nucleotide insertions, deletions, or modifications, such as the conversion of adenosine (A) to inosine (I) or cytidine (C) to uridine (U). RNA editing can occur through different mechanisms and is observed in various organisms, impacting gene expression and protein function.

Chemical modifications of individual nucleotides in RNA, such as methylation, pseudouridylation, and acetylation, are also common. These modifications can influence RNA stability, localization, and interactions with proteins and other molecules, thereby affecting RNA processing and function. Additionally, specific RNA modifications can serve as regulatory marks, influencing RNA-protein interactions and the recognition of specific RNA molecules by cellular machinery (Ferrer-Orta et al., 2006).

KEYWORD

Nucleotide sequence is the most fundamental level of knowledge of a gene or genome. It is the blueprint that contains the instructions for building an organism, and no understanding of genetic function or evolution could be complete without obtaining.

5.3.4. Quality Control Mechanisms in RNA Processing

RNA processing undergoes several quality control mechanisms to ensure the production of functional and accurate RNA molecules. These mechanisms help to maintain the integrity of the RNA transcript and prevent the accumulation of aberrant or defective RNA species. Some key quality control mechanisms involved in RNA processing include (Dynan & Tjian, 1983):

- Splicing Proofreading: The spliceosome, responsible for removing introns and joining exons during splicing, has proofreading mechanisms to ensure accurate splicing. It monitors the splicing reactions and detects errors or inaccuracies. If a mistake is identified, the spliceosome can backtrack and correct the error before proceeding with the splicing process.
- Nonsense-Mediated mRNA Decay (NMD): NMD is a surveillance mechanism that targets and degrades mRNA molecules containing premature termination codons (PTCs) or other defects. PTCs can arise from errors in splicing or mutations in the DNA sequence. NMD helps to prevent the translation of truncated or non-functional

proteins by selectively degrading these faulty transcripts. mRNA Surveillance Pathways: Cells have mechanisms to detect and degrade abnormal or defective mRNAs. These pathways, such as the non-stop decay and no-go decay pathways, recognize stalled or aberrant translation complexes and target the associated mRNA for degradation. These pathways act as quality control mechanisms to ensure the production of high-quality transcripts and prevent the accumulation of potentially harmful or non-functional proteins.

- RNA Editing Quality Control: RNA editing processes, such as adenosine-to-inosine (A-to-I) editing, have built-in quality control mechanisms to ensure accurate editing. Enzymes involved in RNA editing, known as RNA editing enzymes, have proofreading activities that correct errors introduced during the editing process. These proofreading mechanisms help maintain the fidelity of the edited RNA sequences and prevent the generation of transcripts with incorrect edits.

Overall, quality control mechanisms in RNA processing play crucial roles in maintaining the accuracy, integrity, and functionality of RNA molecules. These mechanisms ensure the removal of aberrant transcripts, prevent the production of non-functional proteins, and contribute to the overall precision of gene expression (Tabor & Richardson, 1985).

In summary, RNA processing and maturation involve several important steps, including RNA splicing and alternative splicing, addition of a 5' cap and 3' poly(A) tail, RNA editing and modifications, and quality control mechanisms. RNA splicing removes introns and joins exons to produce mature mRNA molecules, while alternative splicing allows for the generation of multiple mRNA isoforms from a single gene. The addition of a 5' cap and 3' poly(A) tail facilitates mRNA stability, transport, and translation. RNA editing and modifications further expand the diversity and functionality of RNA molecules. Quality control mechanisms ensure the accuracy and integrity of RNA processing, preventing the accumulation of aberrant or defective transcripts. Together, these processes contribute to the production of functional and regulated RNA molecules in cells.

KEYWORD

Genetic code is a set of three-letter combinations of nucleotides called codons, each of which corresponds to a specific amino acid or stop signal.

5.4. TRANSLATION: FROM RNA TO PROTEIN

Translation is a crucial process in gene expression where the genetic information encoded in mRNA is decoded to synthesize proteins. This section will provide an overview of the genetic code, the structure and function of ribosomes, the stages of translation including initiation, elongation, and termination, and the post-translational modifications that proteins undergo.

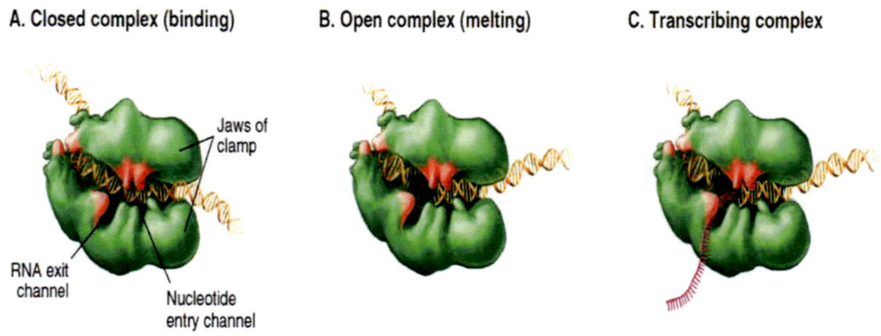

Figure 5.3. RNA polymerase initiation (Source: Abhishek Mazumder, Creative Commons License).

5.4.1. Overview of the Genetic Code

The genetic code is a set of rules that governs the translation of nucleotide sequences in mRNA into the corresponding amino acid sequence of a protein. It is a universal code, meaning that the same codons (three-nucleotide sequences) specify the same amino acids in all living organisms. The genetic code is degenerate, meaning that multiple codons can code for the same amino acid. For example, the amino acid leucine can be specified by six different codons (e.g., UUA, UUG, CUU, CUC, CUA, and CUG). Some amino acids have only one codon (e.g., methionine and tryptophan), while others have multiple codons. In addition to codons that specify amino acids, there are codons that serve as start and stop signals for translation. The start codon, AUG (encoding methionine), signals the initiation of translation, while stop codons (UAA, UAG, and UGA) indicate the termination of translation (Mekler et al., 2002).

5.4.2. Structure and Function of Ribosomes

Ribosomes are large molecular complexes responsible for protein synthesis. They consist of two subunits, a small subunit (40S) and a large subunit (60S) in eukaryotes, or a small subunit (30S) and a large subunit (50S) in prokaryotes. These subunits come together during translation to form a functional ribosome. The small subunit contains the mRNA binding site, while the large subunit contains the aminoacyl-tRNA binding site and the peptidyl transferase center responsible for catalyzing the formation of peptide bonds between amino acids. Ribosomes also contain ribosomal RNA (rRNA) molecules and numerous proteins that provide structural support and enzymatic activities during translation (Kwak et al., 2013).

5.4.3. Initiation, Elongation, and Termination of Translation

Translation occurs in three main stages: initiation, elongation, and termination.

5.4.3.1. Initiation

Initiation begins with the binding of the small ribosomal subunit to the mRNA molecule. The small subunit scans the mRNA until it reaches the start codon (AUG). The start codon is recognized by an initiator tRNA carrying methionine (or formylmethionine in bacteria). The large ribosomal subunit then joins the small subunit, and the initiation complex is formed. Initiation factors assist in this process, ensuring the accuracy and efficiency of translation initiation.

5.4.3.2. Elongation

During elongation, amino acids are added to the growing polypeptide chain. Aminoacyl-tRNA molecules, carrying specific amino acids, enter the ribosome and bind to the codon on the mRNA through complementary base pairing between the codon and the anticodon of tRNA.

TIPS

Utilize interactive animations and models to visualize gene expression processes. Online tools offer engaging visuals for understanding transcription, translation, and regulatory mechanisms. Enhance your understanding by actively exploring these resources and make gene expression come alive through interactive learning.

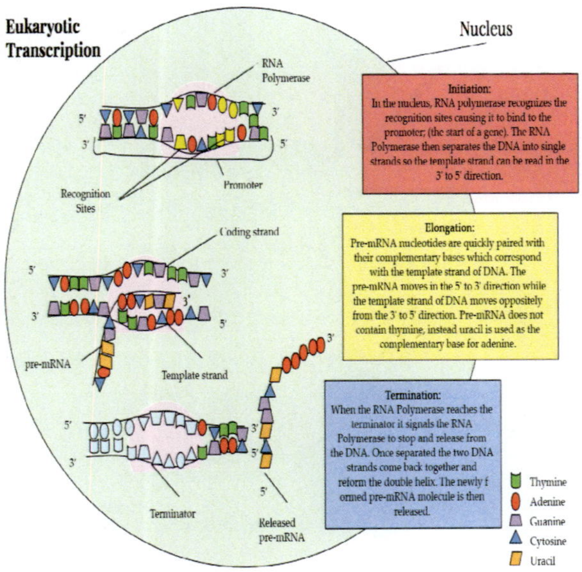

Figure 5.4. Transcription elongation (Source: wikimedia, Creative Commons License).

Peptide bond formation occurs between the amino acids, facilitated by the peptidyl transferase activity of the ribosome. The ribosome then moves along the mRNA in a process known as translocation, shifting the reading frame and exposing the next codon for aminoacyl-tRNA binding. This cycle continues, elongating the polypeptide chain until a stop codon is encountered (Mejía-Almonte et al., 2020).

5.4.3.3. Termination

Termination occurs when a stop codon is reached on the mRNA. Release factors bind to the stop codon, triggering the release of the completed polypeptide chain from the ribosome. The ribosome subunits dissociate, and the mRNA, tRNA, and the newly synthesized protein are released. The polypeptide chain folds into its native three-dimensional structure to become a functional protein (Kaplan et al., 2003).

5.4.4. Post-translational Modifications of Proteins

After translation, proteins often undergo post-translational modifications (PTMs) to acquire their final functional form. These modifications can occur co-translationally or post-translationally and contribute to the diversity and complexity of the proteome. Some common types of PTMs include:

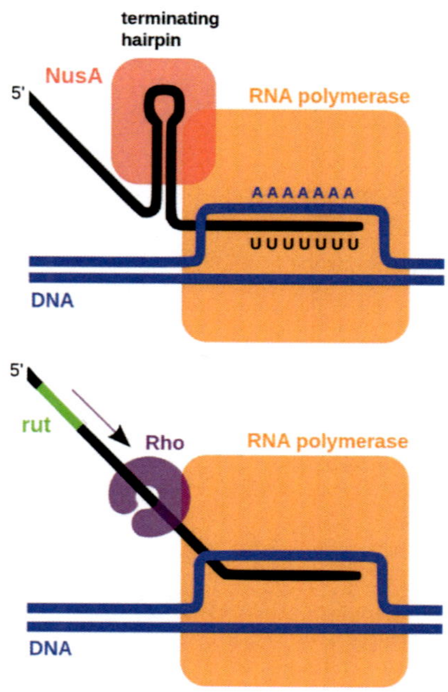

Figure 5.5. Termination of prokaryotic transcription (Source: wikimedia, Creative Commons License).

5.4.4.1. Phosphorylation

Phosphorylation involves the addition of a phosphate group to specific amino acid residues, usually serine, threonine, or tyrosine. Protein kinases catalyze phosphorylation, and protein phosphatases catalyze dephosphorylation. Phosphorylation can modulate protein activity, stability, cellular localization, and protein-protein interactions (Kadonaga, 1998).

KEYWORD

Sequencing DNA means determining the order of the four chemical building blocks - called "bases" - that make up the DNA molecule. The sequence tells scientists the kind of genetic information that is carried in a particular DNA segment.

5.4.4.2. Glycosylation

Glycosylation is the addition of sugar molecules (oligosaccharides) to specific amino acid residues. It can occur on the side chains of asparagine (N-linked glycosylation) or serine/threonine (O-linked glycosylation). Glycosylation affects protein folding, stability, cellular recognition, and function.

5.4.4.3. Acetylation

Acetylation involves the addition of an acetyl group to the amino group of lysine residues. It is mediated by acetyltransferases and removed by deacetylases. Acetylation affects protein-protein interactions, stability, DNA binding, and gene expression regulation (Dynan & Tjian, 1983).

5.4.4.4. Methylation

Methylation involves the addition of a methyl group to specific amino acid residues, such as lysine or arginine. It is catalyzed by methyltransferases and can affect protein-protein interactions, gene expression regulation, and protein localization.

5.4.4.5. Ubiquitination

Ubiquitination is the attachment of ubiquitin molecules to lysine residues in a protein. It is a crucial PTM involved in protein degradation, cellular signaling, and protein trafficking.

These are just a few examples of the diverse array of post-translational modifications that proteins can undergo. PTMs play essential roles in regulating protein function, stability, localization, and interactions, thereby influencing cellular processes, signal transduction pathways, and gene expression (Matsui et al., 1980). In conclusion, translation is the process of converting the genetic information encoded in mRNA into proteins. Ribosomes, composed of rRNA and proteins, perform translation by following the genetic code, where codons in mRNA are matched to specific amino acids. Initiation, elongation, and termination are the stages of translation that involve the binding of ribosomes to mRNA, the addition of amino acids to the growing polypeptide chain, and the recognition of stop codons, respectively. Post-translational modifications further refine and diversify proteins, influencing their activity, stability, localization, and interactions. The accurate and efficient translation and proper post-translational modifications of proteins are essential for cellular function and contribute to the complexity and diversity of the proteome (Neish et al., 1998).

5.5. REGULATION OF GENE EXPRESSION

Gene regulatory elements are specific DNA sequences that control the expression of genes. These elements include promoters, enhancers, silencers, and insulators, which interact with various proteins and regulatory factors to modulate gene expression (Mayer et al., 2010).

5.5.1. Transcription Factors and Their Role in Gene Regulation

Transcription factors are proteins that bind to specific DNA sequences and regulate gene expression.

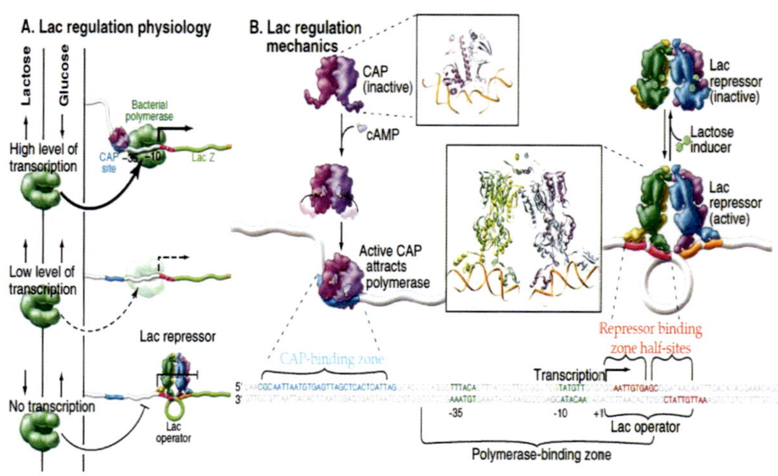

Figure 5.6. Rules of the lactose (lac) operon (Source: Wikimedia, Creative Commons License).

They can activate or repress gene transcription by interacting with other proteins and the transcriptional machinery. Transcription factors play critical roles in development, cellular processes, and response to environmental cues (Turowski & Boguta, 2021).

5.5.2. Chromatin Structure and Epigenetic Modifications

Chromatin structure and epigenetic modifications have profound effects on gene expression. Chromatin, a complex of DNA and proteins, can be tightly packed (heterochromatin) or loosely packed (euchromatin), influencing gene accessibility. Epigenetic modifications, such as DNA methylation and histone modifications, regulate gene expression patterns by altering chromatin structure.

5.5.3. MicroRNAs and Other Non-coding RNAs in Gene Regulation

MicroRNAs (miRNAs) and other non-coding RNAs play crucial roles in post-transcriptional gene regulation. miRNAs bind to target mRNA molecules, leading to their degradation or inhibition of translation. Other non-coding RNAs, such as long non-coding RNAs (lncRNAs), also participate in gene regulation by interacting with proteins or DNA (Arimbasseri et al., 2014). Gene expression is regulated by a complex interplay of gene regulatory elements, transcription factors, chromatin structure, epigenetic modifications, and non-coding RNAs. Understanding these mechanisms is essential for comprehending the precise control of gene expression and its impact on cellular processes and development.

KEYWORD

Gene regulation is the process used to control the timing, location and amount in which genes are expressed. The process can be complicated and is carried out by a variety of mechanisms, including through regulatory proteins and chemical modification of DNA.

5.6. CONTROL OF GENE EXPRESSION

Gene expression is tightly regulated to ensure that genes are expressed in the right cell types, at the right time, and in response to specific signals. The control of gene expression occurs at various levels, including transcriptional regulation, post-transcriptional regulation, and post-translational regulation. This section will provide a detailed explanation of each level of control (Naidu et al., 2011).

5.6.1. Transcriptional Regulation

Transcriptional regulation involves the control of gene expression at the level of transcription initiation. It plays a pivotal role in determining which genes are turned on or off in a given cell or under specific conditions. Two key mechanisms of transcriptional regulation are inducible and repressible operons, and the presence of enhancers, promoters, and silencers.

5.6.1.1. Inducible and Repressible Operons

KEYWORD

Repressible operon is a type of operon in which the transcription of the gene is repressed by the presence of a particular substance. It is regulated by a co-repressor i.e., mainly the end product of the metabolic pathway.

Operons are clusters of genes in bacteria that are transcribed together as a single mRNA molecule. Inducible operons are typically turned on in response to certain environmental cues, such as the presence of specific substrates or inducer molecules. Examples include the lac operon, which is involved in lactose metabolism. Repressible operons, on the other hand, are typically turned off in response to certain molecules or conditions. An example is the trp operon, which is involved in tryptophan synthesis (Gopal et al., 2001).

5.6.1.2. Enhancers, Promoters, and Silencers

Enhancers, promoters, and silencers are DNA sequences that control the transcription of genes. Enhancers are distant regulatory elements that can be located upstream, downstream, or within the gene they regulate. They bind transcription factors and co-activators, which facilitate the recruitment of the transcriptional machinery and enhance gene expression. Promoters are DNA sequences located near the transcription start site and provide binding sites for RNA polymerase and transcription factors. Silencers are DNA sequences that negatively regulate gene expression by binding transcriptional repressors, which inhibit transcription initiation (Lin et al., 2010).

5.6.2. Post-Transcriptional Regulation

Post-transcriptional regulation involves processes that occur after transcription, which impact the stability and processing of mRNA molecules. This level of regulation allows for fine-tuning of gene expression and rapid response to changing conditions. Two important mechanisms of post-transcriptional regulation are RNA stability and degradation, and RNA interference (RNAi) and post-transcriptional gene silencing (PTGS).

5.6.2.1. RNA Stability and Degradation

The stability of mRNA molecules can vary, and it is influenced by sequences in the mRNA itself and by RNA-binding proteins. Specific sequences, such as AU-rich elements (AREs), can target mRNA for rapid degradation. RNA-binding proteins can either stabilize or destabilize mRNA by binding to specific regions. These processes affect the half-life of mRNA, controlling the abundance of specific transcripts and thereby regulating gene expression.

5.6.2.2. RNA Interference and Post-Transcriptional Gene Silencing

RNA interference is a mechanism that regulates gene expression by inhibiting mRNA translation or promoting mRNA degradation. It involves small RNA molecules, such as microRNAs (miRNAs) or small interfering RNAs (siRNAs), which bind to complementary sequences on target mRNA molecules. This binding leads to mRNA degradation or the inhibition of translation, preventing the production of the corresponding protein. This mechanism is essential for fine-tuning gene expression, developmental processes, and defense against viral infections (Missra & Gilmour, 2010).

5.6.3. Post-Translational Regulation

Post-translational regulation involves modifications and processes that occur after protein synthesis. These modifications and processes can impact protein stability, localization, activity, and interactions, providing further control over gene expression. Two important mechanisms of post-translational regulation are protein modification and degradation, and signal transduction pathways that influence gene expression.

5.6.3.1. Protein Modification and Degradation

Proteins can undergo various post-translational modifications, including phosphorylation, acetylation, methylation, ubiquitination, and glycosylation. These modifications can alter protein conformation, activity, stability, and interactions with other molecules. For example, phosphorylation can activate or deactivate proteins by adding or removing phosphate groups. Ubiquitination targets proteins for degradation by the proteasome, controlling their lifespan and abundance (Hayashi et al., 2002).

KEYWORD

Adenylate-uridylate-rich elements are found in the 3' untranslated region (UTR) of many messenger RNAs (mRNAs) that code for proto-oncogenes, nuclear transcription factors, and cytokines.

5.6.3.2. Signal Transduction Pathways and Gene Expression

Signal transduction pathways can influence gene expression by transmitting signals from the cell surface to the nucleus. Activation of cell surface receptors triggers intracellular signaling cascades that ultimately lead to changes in gene expression. These pathways involve various protein kinases, transcription factors, and co-regulators. For example, the mitogen-activated protein kinase (MAPK) pathway can activate specific transcription factors that bind to gene promoters, initiating or enhancing transcription. Post-translational regulation allows for dynamic and rapid adjustments in protein function and abundance. It provides an additional layer of control over gene expression, ensuring that proteins are properly modified, regulated, and degraded to maintain cellular homeostasis and respond to environmental cues (Kuzminov, 2018).

KEYWORD

Mitogen-activated protein kinase is a type of protein kinase that is specific to the amino acids serine and threonine. MAPKs are involved in directing cellular responses to a diverse array of stimuli, such as mitogens, osmotic stress, heat shock and proinflammatory cytokines.

In summary, the control of gene expression occurs at multiple levels, including transcriptional regulation, post-transcriptional regulation, and post-translational regulation. Transcriptional regulation involves the control of gene expression through inducible and repressible operons and the binding of enhancers, promoters, and silencers. Post-transcriptional regulation involves the modulation of mRNA stability and degradation, as well as the regulation of gene expression through RNA interference and post-transcriptional gene silencing. Post-translational regulation involves protein modifications and degradation, as well as the influence of signal transduction pathways on gene expression. These mechanisms work together to ensure precise and dynamic control of gene expression in response to cellular and environmental cues (Kuzminov, 2018).

5.7. TECHNIQUES IN GENE EXPRESSION ANALYSIS

Understanding gene expression patterns and levels is crucial for unraveling biological processes and studying the molecular mechanisms underlying various diseases. This section will provide a detailed explanation of several key techniques used in gene expression analysis, including Northern blotting and RNA quantification, reverse transcription-polymerase chain reaction (RT-PCR), microarray analysis, RNA sequencing (RNA-Seq), gene expression profiling, and the use of bioinformatics tools for data analysis.

5.7.1. Northern Blotting and RNA Quantification

Northern blotting is a classical technique used to detect and quantify specific RNA molecules in a sample. It involves the separation of RNA molecules using gel electrophoresis, followed by their transfer to a membrane for detection. The RNA molecules are then hybridized with labeled probes specific to the RNA of interest. The resulting bands on the membrane can be visualized and quantified to determine the abundance of the target RNA. Although Northern blotting is labor-intensive and limited to analyzing a few genes at a time, it provides information about RNA size and can detect splice variants (Rivetti et al., 2003).

KEYWORD

PCR is a simple and widely used process in which minute amounts of DNA can be amplified into multiple copies. In addition to the rapidity with which this assay works, it is able to quantitatively demonstrate how much of a particular sequence is present.

5.7.2. Reverse Transcription-Polymerase Chain Reaction (RT-PCR)

RT-PCR is a widely used technique for amplifying and quantifying specific RNA molecules in a sample. It involves the reverse transcription of RNA into complementary DNA (cDNA) using reverse transcriptase enzymes, followed by PCR amplification of the cDNA using gene-specific primers. The amplified DNA products can be detected and quantified using various methods, such as gel electrophoresis, fluorescence, or real-time PCR. RT-PCR is sensitive, specific, and allows for the detection and quantification of low abundance RNA molecules. It is commonly used for validating gene expression changes observed in other high-throughput techniques (Simmons et al., 2022).

5.7.3. Microarray Analysis and RNA Sequencing

Microarray analysis and RNA sequencing (RNA-Seq) are high-throughput techniques used to profile gene expression on a genome-wide scale.

5.7.3.1. Microarray Analysis

Microarrays are solid supports, such as glass slides or chips, on which DNA or RNA probes are immobilized in an orderly manner. Sample cDNA or cRNA is labeled and hybridized to the microarray, allowing for the detection and quantification of thousands of genes simultaneously. Microarrays provide information about the expression levels of known genes, and their design can be tailored for specific organisms or gene sets. They have been widely used for gene expression profiling and identifying differentially expressed genes in various biological contexts.

5.7.3.2. RNA Sequencing (RNA-Seq)

RNA-Seq is a next-generation sequencing (NGS) technique that directly sequences cDNA generated from RNA samples. It allows for the identification and quantification of RNA molecules in a sample, including protein-coding mRNA, non-coding RNA, and splice variants. RNA-Seq provides comprehensive gene expression information, detects novel transcripts, and allows for the analysis of alternative splicing, post-transcriptional modifications, and gene fusion events. It has revolutionized the field of transcriptomics and has become the gold standard for gene expression analysis.

KEYWORD

Ingenuity pathway analysis (IPA) is a web-based bioinformatics application that allows researchers to upload data analysis results from high-throughput experiments such as microarray and next generation sequencing for functional analyze, integration, and further understanding.

5.7.4. Gene Expression Profiling and Bioinformatics Tools

Gene expression profiling refers to the measurement of expression levels of many genes simultaneously to gain insights into biological processes. This can be achieved using various techniques mentioned earlier, such as microarrays or RNA-Seq. To analyze and interpret gene expression data, bioinformatics tools and approaches are employed. These tools facilitate data preprocessing, normalization, identification of differentially expressed genes, functional enrichment analysis, pathway analysis, and construction of gene regulatory networks. Examples of widely used bioinformatics tools include R and Bioconductor packages, Gene Set Enrichment Analysis (GSEA), and Ingenuity Pathway Analysis (IPA) (Vincenti & Brinckerhoff, 2002).

5.8. GENE EXPRESSION IN DEVELOPMENT AND DISEASE

Gene expression plays a crucial role in both normal development and disease processes. This section will provide a detailed explanation of the regulation of gene expression during embryonic development, the changes in gene expression observed in cancer and other diseases, and the application of gene therapy for manipulating gene expression.

5.8.1. Regulation of Gene Expression during Embryonic Development

Embryonic development is a highly orchestrated process in which gene expression is tightly regulated to ensure the proper differentiation and organization of cells into tissues and organs. During development, specific genes are turned on or off in a spatial and temporal manner, leading to cell fate determination and the formation of distinct cell types. The regulation of gene expression during embryonic development involves various mechanisms, including the activation or repression of transcription factors, epigenetic modifications, and signaling pathways (Aoyagi & Wassarman, 2000). Transcription factors, such as homeobox genes, play critical roles in specifying cell fate and tissue development. Epigenetic modifications, such as DNA methylation and histone modifications, can silence or activate genes involved in developmental processes. Signaling pathways, such as those mediated by Wnt, Notch, and Sonic Hedgehog, provide spatial and temporal cues that guide gene expression patterns.

Understanding the regulation of gene expression during embryonic development is essential for unraveling the molecular mechanisms underlying organogenesis, tissue patterning, and developmental disorders (So et al., 2007).

ACTIVITY 5.1

Analyze a gene expression dataset comparing cancer cells and normal cells. Identify differentially expressed genes, explore enriched biological pathways, and infer potential therapeutic targets. Discuss validation methods such as RT-PCR or immunohistochemistry to confirm gene expression changes.

5.8.2. Gene Expression Changes in Cancer and Other Diseases

Aberrant gene expression is a hallmark of many diseases, including cancer. In cancer, gene expression profiles are altered, leading to uncontrolled cell growth, invasion, and metastasis. Gene expression changes in cancer can result from genetic mutations, epigenetic modifications, dysregulation of signaling pathways, and altered expression of transcription factors and regulatory proteins. Gene expression profiling techniques, such as microarray analysis and RNA-Seq, have been instrumental in identifying differentially expressed genes in cancer. These studies have revealed specific gene expression signatures associated with different cancer types, subtypes, prognosis, and response to treatment. The identification of key driver genes and dysregulated pathways has led to the development of targeted therapies aimed at restoring normal gene expression patterns or inhibiting the activity of specific oncogenes (Sawadogo & Roeder, 1985).

In addition to cancer, gene expression changes are also observed in various other diseases, including neurodegenerative disorders, cardiovascular diseases, autoimmune diseases, and developmental disorders. Gene expression studies in these diseases have helped identify disease-associated genes, pathways, and potential therapeutic targets.

5.8.3. Gene Therapy and Manipulation of Gene Expression

Gene therapy is an approach aimed at manipulating gene expression to treat genetic disorders, cancer, and other diseases. It involves the delivery of therapeutic genes into cells to restore or modulate gene expression and protein function. Gene therapy strategies can include the introduction of functional genes to replace mutated or missing genes, the use of RNA molecules to interfere with gene expression (RNA interference), or the modification of endogenous genes to correct genetic defects (genome editing). Various delivery methods have been employed for gene therapy, including viral vectors, such as adenoviruses and lentiviruses, and non-viral methods, such as liposomes and nanoparticles. These delivery systems enable the targeted delivery of therapeutic genes to specific cell types or tissues (Landers & Spelsberg, 1992).

KEYWORD

Severe combined immunodeficiency (SCID) is a group of rare disorders caused by mutations in different genes involved in the development and function of infection-fighting immune cells.

Gene therapy has shown promising results in clinical trials for genetic disorders, such as severe combined immunodeficiency (SCID) and certain types of inherited blindness. It also holds potential for the treatment of cancer by delivering therapeutic genes to tumor cells to inhibit their growth or enhance the immune response against them. Manipulation of gene expression through gene therapy provides a promising avenue for treating a wide range of diseases, and ongoing research continues to refine and improve the effectiveness and safety of these approaches (Orphanides et al., 1999).

Gene expression plays a crucial role in both normal development and disease processes. The regulation of gene expression during embryonic development is tightly controlled and involves transcription factors, epigenetic modifications, and signaling pathways. Aberrant gene expression is a common feature of diseases, particularly cancer, where changes in gene expression contribute to the development and progression of the disease. Gene expression profiling techniques have provided insights into disease-specific gene expression signatures, facilitating the identification of potential therapeutic targets. Gene therapy offers a promising approach for manipulating gene expression to treat genetic disorders, cancer, and other diseases. By delivering therapeutic genes or modulating endogenous gene expression, gene therapy holds the potential to restore normal gene expression patterns and alleviate disease symptoms. Continued research and advancements in gene expression analysis and gene therapy hold great promise for improving our understanding of disease mechanisms and developing effective treatments (Cheung et al., 2008).

SUMMARY

The chapter on "Gene Expression" provides an overview of the processes involved in gene expression, its regulation, and its significance in development and disease. It covers transcription and translation, explaining the roles of RNA polymerase, ribosomes, and the genetic code. The chapter also discusses post-transcriptional modifications, gene regulatory elements, chromatin structure, and non-coding RNAs in gene regulation. Techniques such as Northern blotting, RT-PCR, microarray analysis, and RNA sequencing are introduced for gene expression analysis. Additionally, the chapter explores gene expression in embryonic development, disease, and the potential of gene therapy.

REVIEW QUESTIONS

1. Describe the process of transcription, including initiation, elongation, and termination.
2. What are the post-transcriptional modifications that RNA undergoes? Explain their significance.
3. How do enhancers, promoters, and silencers contribute to gene regulation?
4. Discuss the role of microRNAs in post-transcriptional gene regulation.
5. Compare and contrast microarray analysis and RNA sequencing for gene expression studies.
6. Explain the importance of chromatin structure and epigenetic modifications in gene expression regulation.
7. What are the potential applications and challenges of gene therapy in manipulating gene expression for therapeutic purposes?

MULTIPLE CHOICE QUESTIONS

1. Which enzyme is responsible for transcription?

a. DNA polymerase
b. RNA polymerase
c. DNA ligase
d. Ribonuclease

2. What is the function of ribosomes in gene expression?

a. DNA replication
b. RNA splicing
c. Protein synthesis
d. DNA repair

3. **What is the role of transcription factors in gene regulation?**
   a. Enhancing RNA stability
   b. Promoting translation
   c. Initiating transcription
   d. Repairing DNA damage
4. **Which of the following is an example of a post-translational modification?**
   a. mRNA splicing
   b. Addition of a 5' cap
   c. Phosphorylation of a protein
   d. Transcription initiation
5. **What is the purpose of microarray analysis in gene expression studies?**
   a. Detecting mutations in DNA
   b. Quantifying RNA levels
   c. Analyzing protein-protein interactions
   d. Identifying gene regulatory elements
6. **Epigenetic modifications can alter gene expression by:**
   a. Modifying the DNA sequence
   b. Enhancing transcription factor binding
   c. Changing the genetic code
   d. Inhibiting RNA polymerase activity
7. **Gene therapy involves:**
   a. Editing DNA sequences
   b. Analyzing gene expression patterns
   c. Manipulating protein levels
   d. Introducing therapeutic genes

Answers to Multiple Choice Questions

1. (b)	2. (c)	3. (c)	4. (c)	5. (b)
6. (b)	7. (d)			

REFERENCES

1. Aoyagi, N., & Wassarman, D. A. (2000). Genes encoding Drosophila melanogaster RNA polymerase II general transcription factors: diversity in TFIIA and TFIID components contributes to gene-specific transcriptional regulation. *The Journal of Cell Biology*, *150*(2), 45-50.
2. Arimbasseri, A. G., Rijal, K., & Maraia, R. J. (2014). Comparative overview of RNA polymerase II and III transcription cycles, with focus on RNA polymerase III termination and reinitiation. *Transcription*, *5*(1), 27369.
3. Artsimovitch, I., & Landick, R. (2000). Pausing by bacterial RNA polymerase is mediated by mechanistically distinct classes of signals. *Proceedings of the National Academy of Sciences*, *97*(13), 7090-7095.
4. Audic, S., & Claverie, J. M. (1997). The significance of digital gene expression profiles. *Genome research*, *7*(10), 986-995.
5. Barman, B., Sung, B. H., Krystofiak, E., Ping, J., Ramirez, M., Millis, B., ... & Weaver, A. M. (2022). VAP-A and its binding partner CERT drive biogenesis of RNA-containing extracellular vesicles at ER membrane contact sites. *Developmental Cell*, *57*(8), 974-994.
6. Bhalerao, R., Keskitalo, J., Sterky, F., Erlandsson, R., Bjorkbacka, H., Birve, S. J., ... & Jansson, S. (2003). Gene expression in autumn leaves. *Plant physiology*, *131*(2), 430-442.
7. Brazma, A., & Vilo, J. (2000). Gene expression data analysis. *FEBS letters*, *480*(1), 17-24.
8. Brouwer, T. F., Walker, R. G., Chapman, F. W., & Koster, R. W. (2015). Association between chest compression interruptions and clinical outcomes of ventricular fibrillation out-of-hospital cardiac arrest. *Circulation*, *132*(11), 1030-1037.
9. Cheung, V., Chua, G., Batada, N. N., Landry, C. R., Michnick, S. W., Hughes, T. R., & Winston, F. (2008). Chromatin-and transcription-related factors repress transcription from within coding regions throughout the Saccharomyces cerevisiae genome. *PLoS biology*, *6*(11), 277.
10. Coulon, A., Ferguson, M. L., de Turris, V., Palangat, M., Chow, C. C., & Larson, D. R. (2014). Kinetic competition during the transcription cycle results in stochastic RNA processing. *Elife*, *3*, e03939.
11. Davis, S., & Meltzer, P. S. (2007). GEOquery: a bridge between the Gene Expression Omnibus (GEO) and BioConductor. *Bioinformatics*, *23*(14), 1846-1847.
12. Dynan, W. S., & Tjian, R. (1983). Isolation of transcription factors that discriminate between different promoters recognized by RNA polymerase II. *Cell*, *32*(3), 669-680.
13. Edgar, R., Domrachev, M., & Lash, A. E. (2002). Gene Expression Omnibus: NCBI gene expression and hybridization array data repository. *Nucleic acids research*, *30*(1), 207-210.

14. Emilsson, V., Thorleifsson, G., Zhang, B., Leonardson, A. S., Zink, F., Zhu, J., ... & Stefansson, K. (2008). Genetics of gene expression and its effect on disease. *Nature*, *452*(7186), 423-428.

15. Ferrer-Orta, C., Arias, A., Escarmís, C., & Verdaguer, N. (2006). A comparison of viral RNA-dependent RNA polymerases. *Current opinion in structural biology*, *16*(1), 27-34.

16. Fishburn, J., Mohibullah, N., & Hahn, S. (2005). Function of a eukaryotic transcription activator during the transcription cycle. *Molecular cell*, *18*(3), 369-378.

17. Forget, D., Lacombe, A. A., Cloutier, P., Al-Khoury, R., Bouchard, A., Lavallée-Adam, M., ... & Coulombe, B. (2010). The protein interaction network of the human transcription machinery reveals a role for the conserved GTPase RPAP4/GPN1 and microtubule assembly in nuclear import and biogenesis of RNA polymerase II. *Molecular & Cellular Proteomics*, *9*(12), 2827-2839.

18. Goeman, J. J., & Bühlmann, P. (2007). Analyzing gene expression data in terms of gene sets: methodological issues. *Bioinformatics*, *23*(8), 980-987.

19. Gopal, B., Haire, L. F., Gamblin, S. J., Dodson, E. J., Lane, A. N., Papavinasasundaram, K. G., ... & Dodson, G. (2001). Crystal structure of the transcription elongation/anti-termination factor NusA from Mycobacterium tuberculosis at 1.7 Å resolution. *Journal of molecular biology*, *314*(5), 1087-1095.

20. Goverdhana, S., Puntel, M., Xiong, W., Zirger, J. M., Barcia, C., Curtin, J. F., ... & Castro, M. G. (2005). Regulatable gene expression systems for gene therapy applications: progress and future challenges. *Molecular Therapy*, *12*(2), 189-211.

21. Hayashi, T., Takahashi, M., Takaya, Y., & Ogasawara, M. (2002). Catalytic cycle of rhodium-catalyzed asymmetric 1, 4-addition of organoboronic acids. Arylrhodium, oxa-π-allylrhodium, and hydroxorhodium intermediates. *Journal of the American Chemical Society*, *124*(18), 5052-5058.

22. Iyer, L. M., Koonin, E. V., & Aravind, L. (2003). Evolutionary connection between the catalytic subunits of DNA-dependent RNA polymerases and eukaryotic RNA-dependent RNA polymerases and the origin of RNA polymerases. *BMC structural biology*, *3*(1), 1-23.

23. Johnson, P. J., Kooter, J. M., & Borst, P. (1987). Inactivation of transcription by UV irradiation of T. brucei provides evidence for a multicistronic transcription unit including a VSG gene. *Cell*, *51*(2), 273-281.

24. Kadonaga, J. T. (1998). Eukaryotic transcription: an interlaced network of transcription factors and chromatin-modifying machines. *Cell*, *92*(3), 307-313.

25. Kaplan, C. D., Laprade, L., & Winston, F. (2003). Transcription elongation factors repress transcription initiation from cryptic sites. *Science*, *301*(5636), 1096-1099.

26. Koonin, E. V. (1991). The phylogeny of RNA-dependent RNA polymerases of positive-strand RNA viruses. *Journal of General Virology*, *72*(9), 2197-2206.

27. Kuzminov, A. (2018). When DNA topology turns deadly–RNA polymerases dig in their R-loops to stand their ground: new positive and negative (super) twists in the replication–transcription conflict. *Trends in genetics*, *34*(2), 111-120.
28. Kwak, H., Fuda, N. J., Core, L. J., & Lis, J. T. (2013). Precise maps of RNA polymerase reveal how promoters direct initiation and pausing. *Science*, *339*(6122), 950-953.
29. Landers, J. P., & Spelsberg, T. C. (1992). New concepts in steroid hormone action: transcription factors, proto-oncogenes, and the cascade model for steroid regulation of gene expression. *Critical reviews in eukaryotic gene expression*, *2*(1), 19-63.
30. Lane, W. J., & Darst, S. A. (2010). Molecular evolution of multisubunit RNA polymerases: sequence analysis. *Journal of molecular biology*, *395*(4), 671-685.
31. Lewandoski, M. (2001). Conditional control of gene expression in the mouse. *Nature Reviews Genetics*, *2*(10), 743-755.
32. Lin, C., Smith, E. R., Takahashi, H., Lai, K. C., Martin-Brown, S., Florens, L., ... & Shilatifard, A. (2010). AFF4, a component of the ELL/P-TEFb elongation complex and a shared subunit of MLL chimeras, can link transcription elongation to leukemia. *Molecular cell*, *37*(3), 429-437.
33. Lisica, A., Engel, C., Jahnel, M., Roldán, É., Galburt, E. A., Cramer, P., & Grill, S. W. (2016). Mechanisms of backtrack recovery by RNA polymerases I and II. *Proceedings of the National Academy of Sciences*, *113*(11), 2946-2951.
34. Matsui, T., Segall, J., Weil, P. A., & Roeder, R. G. (1980). Multiple factors required for accurate initiation of transcription by purified RNA polymerase II. *Journal of Biological Chemistry*, *255*(24), 11992-11996.
35. Mayer, A., Lidschreiber, M., Siebert, M., Leike, K., Söding, J., & Cramer, P. (2010). Uniform transitions of the general RNA polymerase II transcription complex. *Nature structural & molecular biology*, *17*(10), 1272-1278.
36. Mejía-Almonte, C., Busby, S. J., Wade, J. T., van Helden, J., Arkin, A. P., Stormo, G. D., ... & Collado-Vides, J. (2020). Redefining fundamental concepts of transcription initiation in bacteria. *Nature Reviews Genetics*, *21*(11), 699-714.
37. Mekler, V., Kortkhonjia, E., Mukhopadhyay, J., Knight, J., Revyakin, A., Kapanidis, A. N., ... & Ebright, R. H. (2002). Structural organization of bacterial RNA polymerase holoenzyme and the RNA polymerase-promoter open complex. *Cell*, *108*(5), 599-614.
38. Missra, A., & Gilmour, D. S. (2010). Interactions between DSIF (DRB sensitivity inducing factor), NELF (negative elongation factor), and the Drosophila RNA polymerase II transcription elongation complex. *Proceedings of the National Academy of Sciences*, *107*(25), 11301-11306.
39. Mohr, G., Silas, S., Stamos, J. L., Makarova, K. S., Markham, L. M., Yao, J., ... & Lambowitz, A. M. (2018). A reverse transcriptase-Cas1 fusion protein contains a Cas6 domain required for both CRISPR RNA biogenesis and RNA spacer acquisition. *Molecular cell*, *72*(4), 700-714.
40. Naidu, S., Friedrich, J. K., Russell, J., & Zomerdijk, J. C. (2011). TAF1B is a

TFIIB-like component of the basal transcription machinery for RNA polymerase I. *Science*, *333*(6049), 1640-1642.

41. Neish, A. S., Anderson, S. F., Schlegel, B. P., Wei, W., & Parvin, J. D. (1998). Factors associated with the mammalian RNA polymerase II holoenzyme. *Nucleic acids research*, *26*(3), 847-853.
42. O'Halloran, T. V. (1993). Transition metals in control of gene expression. *Science*, *261*(5122), 715-725.
43. Orphanides, G., Wu, W. H., Lane, W. S., Hampsey, M., & Reinberg, D. (1999). The chromatin-specific transcription elongation factor FACT comprises human SPT16 and SSRP1 proteins. *Nature*, *400*(6741), 284-288.
44. Qian, X., Gozani, S. N., Yoon, H., Jeon, C., Agarwal, K., & Weiss, M. A. (1993). Novel zinc finger motif in the basal transcriptional machinery: three-dimensional NMR studies of the nucleic acid binding domain of transcriptional elongation factor TFIIS. *Biochemistry*, *32*(38), 9944-9959.
45. Rivetti, C., Codeluppi, S., Dieci, G., & Bustamante, C. (2003). Visualizing RNA extrusion and DNA wrapping in transcription elongation complexes of bacterial and eukaryotic RNA polymerases. *Journal of molecular biology*, *326*(5), 1413-1426.
46. Robinson, M. D., McCarthy, D. J., & Smyth, G. K. (2010). edgeR: a Bioconductor package for differential expression analysis of digital gene expression data. *bioinformatics*, *26*(1), 139-140.
47. Romero, I. G., Ruvinsky, I., & Gilad, Y. (2012). Comparative studies of gene expression and the evolution of gene regulation. *Nature Reviews Genetics*, *13*(7), 505-516.
48. Sawadogo, M., & Roeder, R. G. (1985). Interaction of a gene-specific transcription factor with the adenovirus major late promoter upstream of the TATA box region. *Cell*, *43*(1), 165-175.
49. Schadt, E. E., Monks, S. A., Drake, T. A., Lusis, A. J., Che, N., Colinayo, V., ... & Friend, S. H. (2003). Genetics of gene expression surveyed in maize, mouse and man. *Nature*, *422*(6929), 297-302.
50. Schneider, D. A. (2012). RNA polymerase I activity is regulated at multiple steps in the transcription cycle: recent insights into factors that influence transcription elongation. *Gene*, *493*(2), 176-184.
51. Simmons, T. R., Ellington, A. D., & Contreras, L. M. (2022). RNP-Based Control Systems for Genetic Circuits in Synthetic Biology Beyond CRISPR. *Methods in molecular biology (Clifton, NJ)*, *2518*, 1-31.
52. So, A. Y. L., Chaivorapol, C., Bolton, E. C., Li, H., & Yamamoto, K. R. (2007). Determinants of cell-and gene-specific transcriptional regulation by the glucocorticoid receptor. *PLoS genetics*, *3*(6), 94.
53. Sparano, J. A., Gray, R. J., Makower, D. F., Pritchard, K. I., Albain, K. S., Hayes, D. F., ... & Sledge, G. W. (2015). Prospective validation of a 21-gene expression assay in breast cancer. *New England Journal of Medicine*, *373*(21), 2005-2014.

54. Sun, R. E. N., Lin, S. F., Staskus, K., Gradoville, L., Grogan, E., Haase, A., & Miller, G. (1999). Kinetics of Kaposi's sarcoma-associated herpesvirus gene expression. *Journal of virology*, *73*(3), 2232-2242.
55. Tabor, S., & Richardson, C. C. (1985). A bacteriophage T7 RNA polymerase/promoter system for controlled exclusive expression of specific genes. *Proceedings of the National Academy of Sciences*, *82*(4), 1074-1078.
56. Thiffault, I., Wolf, N. I., Forget, D., Guerrero, K., Tran, L. T., Choquet, K., ... & Bernard, G. (2015). Recessive mutations in POLR1C cause a leukodystrophy by impairing biogenesis of RNA polymerase III. *Nature communications*, *6*(1), 1-9.
57. Turowski, T. W., & Boguta, M. (2021). Specific Features of RNA Polymerases I and III: Structure and Assembly. *Frontiers in Molecular Biosciences*, *8*, 680090.
58. Venkataraman, S., Prasad, B. V., & Selvarajan, R. (2018). RNA dependent RNA polymerases: insights from structure, function and evolution. *Viruses*, *10*(2), 76.
59. Vincenti, M. P., & Brinckerhoff, C. E. (2002). Transcriptional regulation of collagenase (MMP-1, MMP-13) genes in arthritis: integration of complex signaling pathways for the recruitment of gene-specific transcription factors. *Arthritis research & therapy*, *4*(3), 1-8.
60. Wandzik, J. M., Kouba, T., Karuppasamy, M., Pflug, A., Drncova, P., Provaznik, J. ... & Cusack, S. (2020). A structure-based model for the complete transcription cycle of influenza polymerase. *Cell*, *181*(4), 877-893.
61. White, R. J. (2005). RNA polymerases I and III, growth control and cancer. *Nature reviews Molecular cell biology*, *6*(1), 69-78.
62. Wingender, E., Chen, X., Fricke, E., Geffers, R., Hehl, R., Liebich, I., & Urbach, S. (2001). The TRANSFAC system on gene expression regulation. *Nucleic acids research*, *29*(1), 281-283.
63. Zhang, L., Zhou, W., Velculescu, V. E., Kern, S. E., Hruban, R. H., Hamilton, S. R., ... & Kinzler, K. W. (1997). Gene expression profiles in normal and cancer cells. *Science*, *276*(5316), 1268-1272.
64. Zieg, J., Silverman, M., Hilmen, M., & Simon, M. (1977). Recombinational switch for gene expression. *Science*, *196*(4286), 170-172.

CHAPTER 6

DNA AND RNA

INTRODUCTION

Except for certain viruses, the genetic code for life is written in a DNA molecule (aka deoxyribonucleic acid). No human language can compare to the simplicity & sophistication of DNA in terms of design. DNA is unparalleled in this regard. DNA, on the other hand, is a linguist's worst nightmare from the point of view of implementation, which refers to how It is written & spoken in practice (Joachimi, Benz, & Hartig, 2009). There are four primary roles that DNA plays:

- It carries the instructions needed to make proteins and enzymes;
- It is involved in the regulation of the times at which proteins and enzymes are synthesized as well as the times at which they are not made;
- As cells divide, it transmits this information to the daughter cells;
- This information is passed on from parental organisms to their offspring by this mechanism (Collie & Parkinson, 2011).

In this chapter, we'll investigate its structure, language, & process by which the genetic template is converted into a protein that can be used by the cell.

Learning Objectives

At the end of this chapter, the reader will be able to understand

- The physical structure of DNA
- The importance of DNA for the development of humans
- The concept of RNA and the genetic code
- The process of protein synthesis
- The case study of hemoglobin chain

Key Terms

- Deoxyribonucleic Acid (DNA)
- Ribonucleic acid
- Genetic Code
- Ribosomal RNA
- Polypeptide Chain
- Protein Synthesis
- Transfer RNA

6.1. PHYSICAL STRUCTURE OF DNA

Very few persons in civilizations where literacy is prevalent can avoid seeing a picture of DNA.

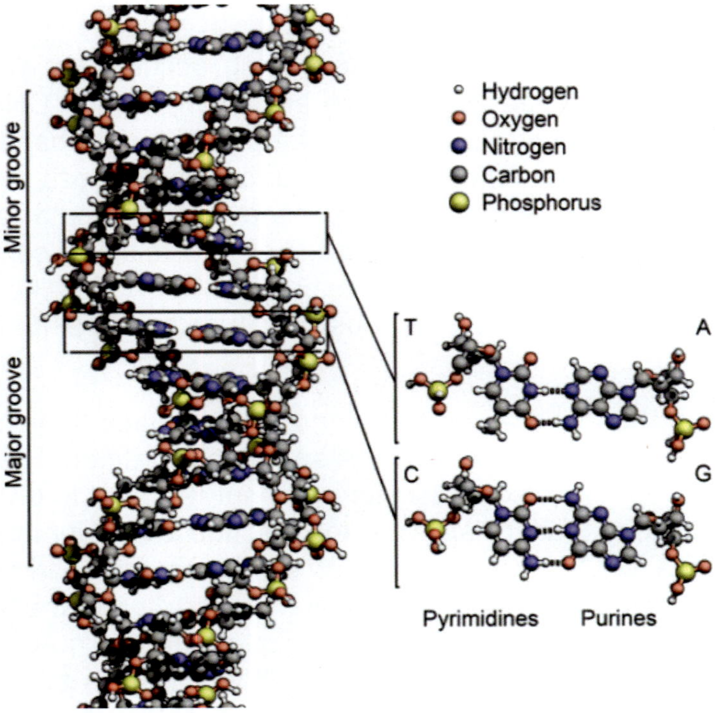

Figure 6.1. Illustration of the structure of DNA (Source: Wikimedia, Creative Commons License).

The structure of DNA is very similar to that of a spiral staircase. For this discussion, let us pretend that we rotate the stairwell to remove the spiral, leaving us with the structure resembling a ladder shown in Figure 6.1. Sugars and phosphates make up the two backbones of this ladder; however, we no longer need to worry about them. The rungs are where all of the action in DNA takes place (Watson & Crick, 1953).

Each rung of the ladder is made up of two different substances chemically bound to each other and referred to as nucleotides or base pairs. Adenine, guanine, thymine, & cytosine are four nucleotides that makeup DNA. These nucleotides are commonly shortened by the initial letter of their names: A, T, G, & C, respectively. DNA contains no other nucleotides. Because of the significance of these four nucleotides, you should make an effort to commit their names to memory (Geider & Hoffmann-Berling, 1981).

Examining Figure 6.1 demonstrates that the nucleotides do not form pairs with one another in a completely random manner. Instead, there is a strong likelihood that A will go with T, and there is a certainty that G will go with C. This fundamental concept behind complementary base pairing is vital for knowing many aspects of the functional behavior of DNA. This is because complementary base pairing is the mechanism by which DNA copies itself (Zimmerman, 1982).

Because complementary base pairing occurs, we will always be aware of the other strand if we are aware of one DNA strand (or helix). Imagine that we removed an entire right-hand side of the DNA ladder from Figure 6.1 and cut it in half across the middle of each rung. The complementary base pairing would still allow us to determine the nucleotide order of this missing component. For example, if the left-hand side begins with ATGCTC, the one that is on the right-hand side must begin with TACGAG (Choo & Klug, 1997).

KEYWORD

Nucleotide is one of the structural components, or building blocks, of DNA and RNA. A nucleotide consists of a base (one of four chemicals: adenine, thymine, guanine, and cytosine) plus a molecule of sugar and one of phosphoric acid.

DNA comes with a certain orientation in space, which causes the "top" and "bottom" of the DNA sequence to differ. The causes of this are too complex to discuss here, but It is vital to understand the terminology geneticists use to describe the orientation. The 5' end of a DNA sequence, which stands for "five prime," is at the top, while the 3' end, which stands for "three prime," is at the bottom. If DNA nucleotide sequence number 1 may be found between DNA sequence number 2 & "top," then it is considered to be *upstream* of DNA sequence number 2. It is *downstream* of sequence 2 if located between sequence 2 & 3' end (Zwolak & Di Ventra, 2008).

6.2. DNA REPLICATION

Moreover, complementary base pairing helps with DNA replication, which is the accurate replication of the DNA sequence. Both daughter cells produced after a cell divide must have the same genetic material. As a result, DNA must be copied twice, one copy going into one cell and the other into the second. The replication procedure must be performed, and it must be done so with a high level of fidelity (Burgers, 1998). Neurons are a prominent exception to the rule that most of our body's cells constantly divide and replace themselves. For instance, some skin cells have an average lifespan of one to two days. Hence, the skin that you & I had a month ago is not the same as the skin which we've now. By the time we reach the age of 80, our skin will have gone through more than 10,000 generations (Matthew Meselson & F Wo Stahl, 1958).

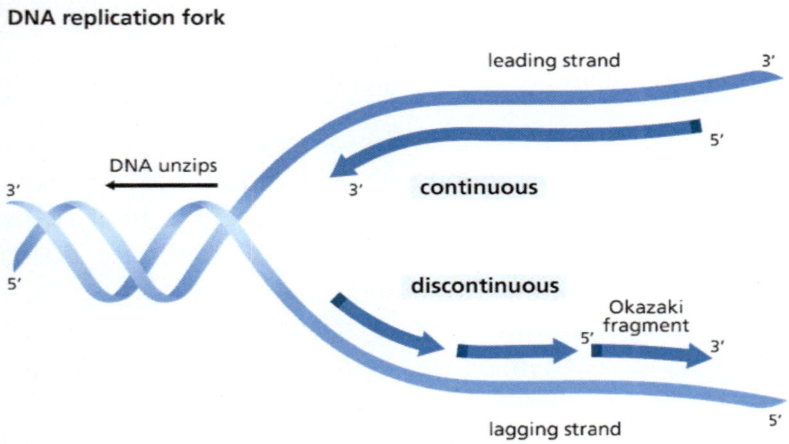

Figure 6.2. Schematic of DNA replication (Source: Your genome, Creative Commons License).

The process of replication is dependent on a chain of proteins and enzymes, which we will refer to as replication material. An enzyme (we can't get away from those enzymes, can we?) initiates the first phase of DNA replication. Divides the rungs like our legendary saw did by splitting them along the middle. Then, enzymes obtain nucleotides floating freely in a cell, attach them to the correct partners on separated stands, & create a new backbone (Wigler, Levy, & Perucho, 1981). The situation is comparable to opening your coat's zipper while new teeth form as the old teeth separate. The original zipper's left side's liberated teeth are bound by one set of new teeth, while another set of new teeth binds the original right side's teeth. Two fully closed zippers remain at the bottom, one on the left & other on the right side of your jacket's front (Matthew Meselson & Franklin W Stahl, 1958).

Did you Know?

DNA polymerase, the enzyme responsible for replicating DNA, can make mistakes during replication, resulting in mutations that can have serious consequences such as cancer.

6.3. RNA

Table 6.1. Tabular representation of some important types of RNA (Source: Your genome, Creative Commons License)

Name	Abbreviation	Function
Messenger RNA	mRNA	Carries the message from the DNA to the protein factory
Ribosomal RNA	rRNA	Comprises part of the protein factory

Transfer RNA	tRNA	Transfers the correct building block to the nascent protein
Interfering RNA	iRNA	Interferes with the DNA message

Before moving on to explain the primary purpose of DNA, it is necessary to first discuss ribonucleic acid, also known as RNA. RNA is DNA's first cousin. RNA and DNA, in addition to their respective chemical compositions, share a number of important similarities and differences. To begin, RNA, like DNA, is comprised of a total of four and only four nucleotides. Nevertheless, in contrast to DNA, RNA replaces thymine (T) with uracil, which is denoted by the letter U. As a consequence of this, the four nucleotides that make up RNA are adenine (A), cytosine (C), guanine (G), & uracil (U) (Cleaver, 1968).

Figure 6.3. Image of RNA (Source: Your Genome, Creative Commons License).

Second, complementary base pairing can be found in RNA's nucleotides as well. DNA and other types of RNA molecules are capable of forming bonds with the RNA nucleotides. Both RNA and DNA will always pair G & C together, but in RNA, T will pair with A whereas DNA will always pair A with U. G & C will always pair together in DNA. When RNA combines with itself, G forms a pairing with C, while A combines with U (Räschle et al., 2008).

DNA is comprised of two strands, whereas RNA is more frequently composed of a single strand. In other words, RNA doesn't have a ladder-like structure that DNA does. Instead, RNA would look like when the ladder was cut in half and the unused portion was taken away (with the added provison that U would substitute for T in the remaining half) (Hurwitz, Dean, Kwong, & Lee, 1990).

Fourth, while there is only one kind of DNA, there are many different kinds of RNA, and each of these RNA forms performs a different function inside the cell. In Table 6.1, the many categories of importance for this function are listed. Take note of the acronyms. Think of DNA as the cell's king, who commands it to do its functions. But unlike human kings, king DNA can never carry out his orders since he cannot leave the cell's nucleus, where his throne is. The various varieties of RNA represent the numerous henchmen that carry out the King's orders (Harland & Laskey, 1980). Ribosomal RNA is found in the specialized nucleolus in the nucleus. At the same time, transfer RNA transports materials to important sites, and still, other types serve as messengers that carry synthesis instructions (messenger RNA). Interfering RNA, a fourth type, actively obstructs other monarchy messages! The genetic code, which communicates through complementary base pairing, is the universal language of the universe, as we shall show (López-Contreras & Fernandez-Capetillo, 2010).

Remember

The genetic code is the set of rules by which information encoded within DNA is translated into proteins. While the code itself is universal across all living organisms, the ways in which it is regulated and expressed can vary widely. Understanding the genetic code is essential for understanding how organisms develop, grow, and function.

6.4. GENETIC CODE

It is written in the DNA. It doesn't build anything. It is vital to understand the genetic code before explaining how information in DNA leads to the production of specific molecules. It is convenient to think of each species' genome as a book, with the genetic code serving as a universal language of all books in the universe. This language's "alphabet" consists of four letters, represented by the nucleotides A, T, C, & G in DNA or RNA (A, U, G, & C) (Wang & Schultz, 2005).

A genetic "word" is termed a codon and, unlike words in human language, which may contain any number of letters, only three nucleotides make up a codon. Every genetic term represents an amino acid. (An amino acid will be defined further.) For instance, amino acid phenylalanine is "DNAese" for nucleotide sequence AAG. The amino acid glutamine is "DNAese" for nucleotide sequence GTC (Francis HC Crick, 1968).

DNA has synonyms, much like natural language does. In other words, different triplet nucleotide sequences can represent the same amino acid. For instance, DNA codes ATA and ATG stand for the amino acid tyrosine (Nirenberg, 1963).

A string of words that specify an amino acid sequence makes up a sentence in the DNA language. The amino acids leucine, histidine, serine, and valine make up the polypeptide chain represented by the DNA sequence AACGTATCGCAT. There is no need for gaps between the words because the DNA language is triplet-based. If you start from the right place, the language will translate accurately (Wang, Xie, & Schultz, 2006).

Punctuation is an element of the DNA language, just like in natural written language. For instance, nucleotide DNA triplets ATT, ATC, & ACT are comparable to the period (.) used to conclude sentences since they all signify the conclusion of a polypeptide chain. *Stop codons* are what they are termed. A *start codon* is TAC. It serves as a capital letter signifying the beginning of a new sentence. It tells a cell to "start peptide chain here" in a biological sense (Francis HC Crick, Barnett, Brenner, & Watts-Tobin, 1961).

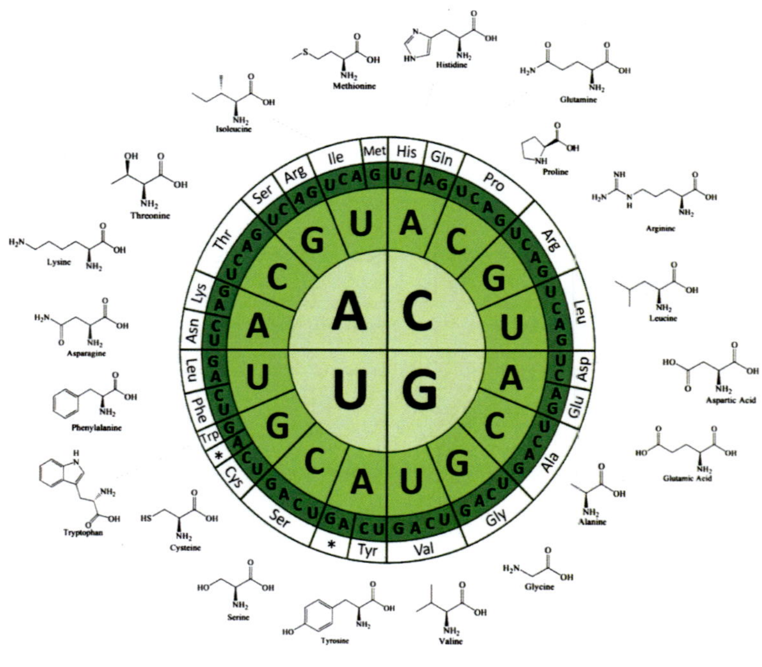

Figure 6.4. Image of the genetic code (Source: Fages-Lartaud et al., Creative Commons License).

Ultimately, DNA is divided into chapters, exactly like a book. The number of chapters varies from one species to the next because they correlate to the chromosomes. There are 24 different chapters in the book for humans, one for each chromosome (22 autosomes plus the X and the Y chromosomes). There is a minimal association between the complexity of the living forms & number of chapters, so the book for other species may have fewer or more chapters (Freeland & Hurst, 1998).

The discrepancies are as significant as the similarities between DNAese and real human language. The only difference between human language and DNA is that the former is coherent, while the latter is the most cluttered & disjointed communication system ever synthesized. A human language book's chapters are first structured in a logical order to tell a tale. The chromosomes are not arranged in such a way (Woese, 1965).

KEYWORD

Deoxyribonucleic acid (DNA) is the molecule that carries genetic information for the development and functioning of an organism.

Table 6.2. A representation of the genetic code in DNA as a wheel (Source: Wikimedia, Creative Commons License)

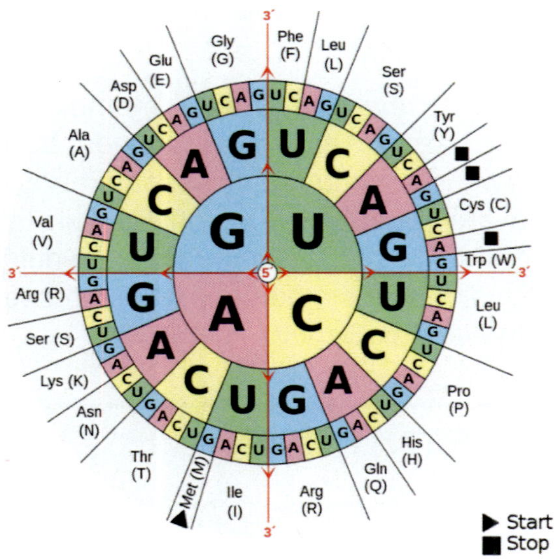

Amino Acids (with single letter abbreviation): Ala = Alanine(A); Arg = Arginine(R); Asn = Asparagine(N); Cys = Cysteine(C); Asp = Aspartic acid(D); Gln = Glutamine(Q); Gly = Glycine(G); Glu = Glutamic Acis(E); His = Histidine(H); Ile = Isoleucine(I); Leu = Leucine(L); Lys = Lysine (K); Met = Methionine(M); Phe = Phenylalanine(P); The = Threonine (T); Trp = Tryptophan(W); Val = Valine(V); Ser = Serine(S); Tyr = Tyrosine(Y)

Second, English phrases physically follow one another, with one sentence defining, elaborating, or providing additional details to another to finish the notion. Seldom, if ever, does the genetic language follow a rational progression. In a metaphorical sense, the first DNA sentence may be a weather forecast, the second might list the components for a chili recipe, & third might be a wise saying about politics thirdly, even though it is ridiculous to compose a compound sentence in English with a paragraph or two in between two separate clauses. Independent clauses of the same sentence are regularly inserted into different chapters of DNA (Francis HC Crick, 1966).

KEYWORD

Serine is an amino acid. It comes in two forms: L-serine and D-serine. L-serine is consumed in the diet and D-serine is made in the body from L-serine.

Fourth, no English-language book would be released where most sentences are broken up by what appears to be a chimpanzee aimlessly tapping a keyboard. Over dozen lengthy strings of this obvious gibberish can puncture a single DNA phrase. Fifth, using the same idea in two phrases simultaneously in natural language is regarded as poor rhetoric. Repetition frequently occurs in DNA, but not seldom. In addition to having nonsensical passages repeated thousands of times, often in the same chapter, DNA has a constant tendency to stutter, stammer, and hem and haw (Wang, Brock, Herberich, & Schultz, 2001).

Last but not least, the size of each mammalian species' DNA "book" dwarfs the size of any book penned by a human. In a 500-page book with thirty or so lines and about 80 characters each line, there are roughly 1,500,000 English letters. To fit the humanDNA book inside one of these books would require nearly 2,000. Also, the majority of the personalities in these 2,000 volumes are mysterious! (Wang & Schultz, 2002).

6.5. PROTEIN SYNTHESIS

6.5.1. Definitions

We'll now look at the specifics of how the DNA template directs protein production. Although we have already discussed proteins and enzymes, we must look more closely at these substances. Amino acid is the fundamental component of every protein and enzyme. Twenty amino acids are required to build proteins, most of which have suffix "ine," such as phenylalanine, serine, and tyrosine. The first three letters of an amino acid's name are typically used to abbreviate it, such as phe for phenylalanine and tyr for tyrosine (Carter & Houlihan, 2001). The 3 main sources of amino acids in the human body are as follows. First, amino acids can be synthesized by our bodies' cells using other, more basic substances (or maybe from other amino acids). Second, amino acids are continuously formed from cell proteins and enzymes. Finally, our diets can provide us with amino acids. Enzymes in our intestine & stomach convert the protein in a juicy steak into amino acids when consumed. The blood then carries these amino acids to various bodily cells (Francis H Crick, 1958).

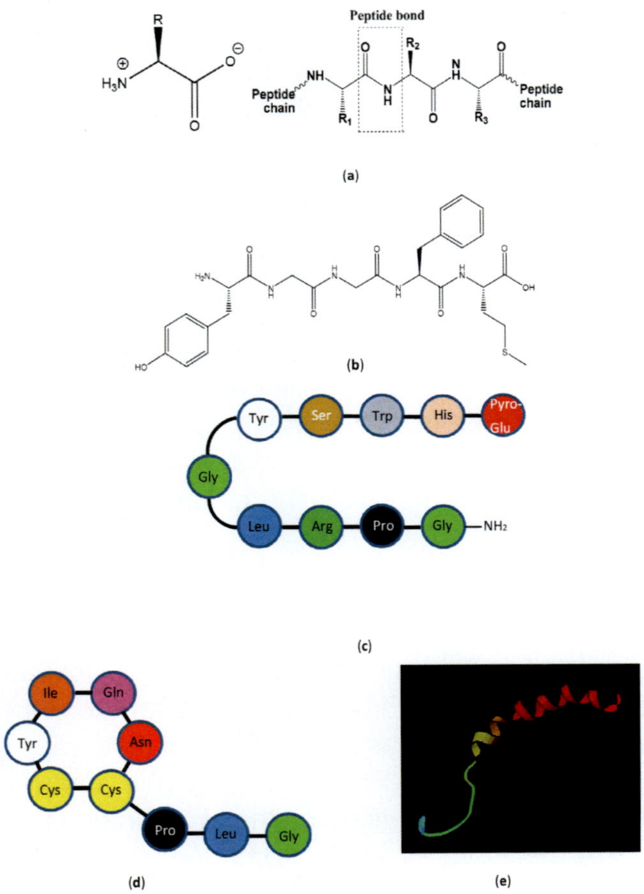

Figure 6.5. Image of a polypeptide chain (Source: Trier et al., Creative Commons License).

A *polypeptide chain* is a collection of amino acids that are physically connected. For now, picture a polypeptide chain as a linear collection of connected boxcars. The boxcars and their couplings represent the chemical bonds keeping the amino acids together. Because it doesn't branch into a Y-shaped structure, the series is linear. The concept of a polypeptide chain is vitally essential for a comprehensive comprehension of genes, so give us some leeway to ramble on about nomenclature (Haselkorn & Rothman-Denes, 1973).

TIPS

Remember that the process of protein synthesis is highly regulated and involves many different steps. Even a small mistake in this process can have serious consequences for the body. Pay attention to your protein intake and make sure you are getting enough of the essential amino acids.

However, the vocabulary employed to describe polypeptide chains lacks formal rules, making it difficult for beginners to understand. Whenever there are less than twelve amino acids in a chain of linked amino acids, the term "peptide" is typically used to characterize the structure. In addition, the term "peptide" can be used both as an adjective and as a suffix to refer to something that is constructed out of amino acids. For example, neuropeptide is a sequence of linked amino acids that can be found in a neuron, & peptide hormone is a hormone that is composed of linked amino acids. Both of these types of peptides are found in neurons. A larger series of connected amino acids, often numbering in the thousands, are typically called *polypeptide chains or polypeptides*. Be cautious, though. There are always going to be exclusions to this usage (Moldave, 1985).

We're now prepared to define our trusty protein. One or more polypeptide chains that are physically linked and assume a three-dimensional structure are known as proteins. A protein comprises one or more polypeptide chains that will bend, fold back on themselves, & connect at different locations to form a molecule that's not just a simple linear structure. Hemoglobin, a protein found in red blood cells which transports oxygen, serves as an example. Four polypeptide chains make up its structure; they fold, connect, and bond (Houlihan, Carter, & McCarthy, 1995).

Other than amino acids, certain proteins also contain additional substances. For instance, a lipoprotein also contains a lipid, or fat, in addition to an amino acid chain. Many receptors found on cell membranes (but occasionally inside cells) are complexes of several proteins and lipids. Finally, It is important to remember what an enzyme is. A specific class of protein tasked with metabolism is known as an enzyme (Proud, 2006). We can now provide one definition of a gene in light of these definitions. A gene is a *section of DNA that houses instructions needed to make a peptide or polypeptide chain*. These genes are occasionally described as *structural genes or coding areas*. The Latin word *locus*, which translates to "site, place, or placement," is a synonym for the gene (plural: loci) (Wilken & Kent, 1998).

6.5.2. Process of Protein Synthesis

Now that the genetic template has been translated into a physical molecule, we can examine the actual "synthesis process." This technique consists of five steps. We will go through each of these individually (Palade, 1975).

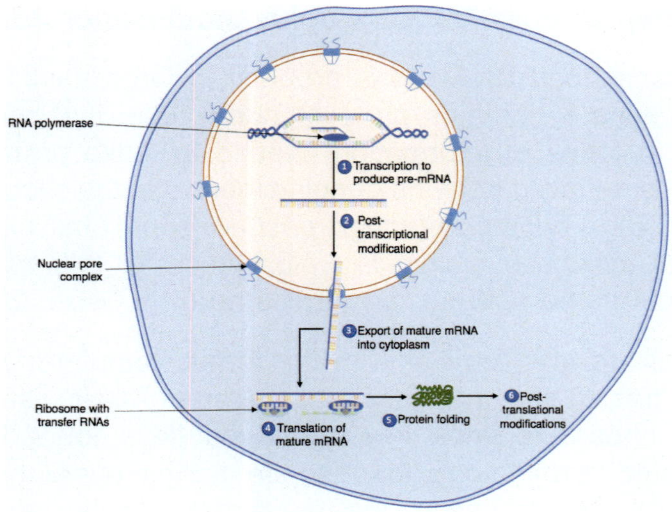

Figure 6.6. Illustration of protein synthesis (Source: Wikimedia , Creative Commons License).

6.5.2.1. Step 1: Transcription

Transcription is a process by which a portion of DNA is "read" & chemically "photocopied" into an RNA molecule.

Similar to DNA replication, transcription requires various enzymes, each carrying out a specific function, like unwinding the double helix, breaking links to make DNA single-stranded, or adding RNA nucleotides. Let's refer to these enzymes as "*transcription things*" and ignore their names and functions. There is a promoter region in the DNA that triggers transcription. To begin the transcription process, the transcription machinery binds to and recognizes a piece of DNA known as a *promoter region*. Just before the coding region, at the start of the gene, is the promoter region (the region of nucleotides within the gene that specifies how proteins should be constructed) (Hager, McNally, & Misteli, 2009)

The transcription material binds to the DNA, unwinds the double helix, & breaks connections between nucleotides to turn DNA into a single strand. The enzymes then create an RNA molecule from a DNA strand. Protein synthesis is accomplished through transcription (Latchman, 1997)

That involves snatching up free nucleotides, choosing one that matches the DNA nucleotide that is currently present, and "gluing" it to the RNA chain. A synthesized RNA sequence will read CGAUCU if the DNA sequence is GCTAGA. This process correctly preserves information in DNA & RNA, although as a genetic "mirror image."(Bucholtz, 2000)

6.5.2.2. Step 2: Post-Transcriptional Modification (Editing)

Freshly transcribed DNA has a problem in large multicellular life forms because it contains parts of nucleotides that are useless gibberish in terms of information needed to make protein. When nonsense is removed from the genuine message, the second protein synthesis stage occurs. Although *post-transcriptional modification* is the term molecular biologists typically use for this step, *editing* is its logical name (Oerum, Dégut, Barraud, & Tisné, 2017).

KEYWORD

Cytoplasm is the gelatinous liquid that fills the inside of a cell. It is composed of water, salts, and various organic molecules.

You all are also familiar with the terms "template," "message section," "trash," and "non-message segment." Scientists find this scenario unbearable since they spend countless hours developing sophisticated terminology to describe things everyone already understands. Post-transcriptional modification is the same. *Exons* are the regions of RNA that hold the real genetic information and template for the polypeptide chain. *Introns* are the parts of transcribed RNA which do not have any message (Nachtergaele & He, 2017).

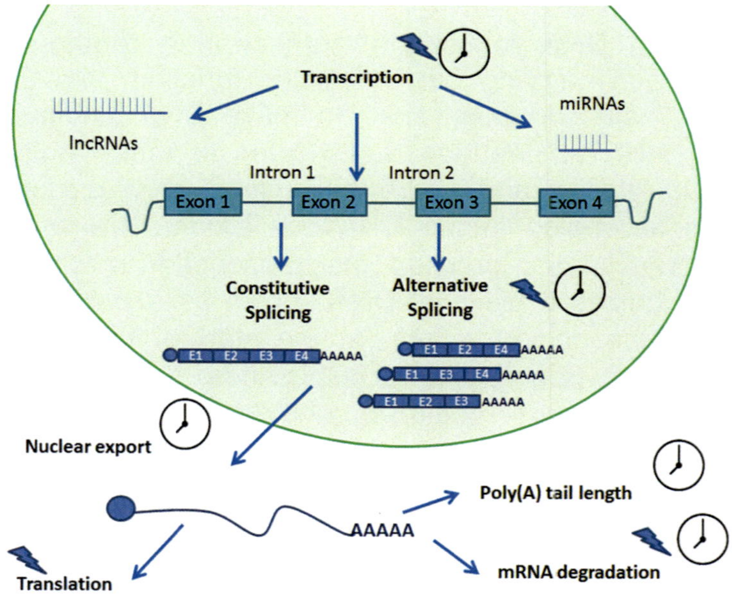

Figure 6.7. Schematic of post-transcriptional modification (editing): Protein synthesis (Source: Romanowski A & Yanovsky MJ, Creative Commons License).

As a result, during editing, the exons are joined together, and the introns are removed. The resulting molecule is referred to as *messenger RNA* or *mRNA*. The actual coding for the peptide chain follows some "header information" at the beginning of messenger

RNA. Moreover, the molecule has "tail information" in the form of RNA nucleotides connected to the end (Kiss, 2001).

Codon is a key concept in *mRNA*. Three consecutive mRNA nucleotides form a codon, where information for a particular amino acid is found. (The triplet in DNA that gives rise to the three nucleotides in mRNA is sometimes called a codon).

6.5.2.3. Step 3: Transportation

In the nucleus, transcription and editing occur. But ribosomes, which are the cell's protein makers, are outside the nucleus in the cytoplasm. The following stage in the production of proteins is straightforward: the mRNA leaves the nucleus, enters the cytoplasm, & connects to a ribosome using its RNA "header information." *Transportation* is the name of this action (Machnicka, Olchowik, Grosjean, & Bujnicki, 2014).

KEYWORD

Messenger RNA (mRNA) is a type of single-stranded RNA involved in protein synthesis. mRNA is made from a DNA template during the process of transcription.

6.5.2.4. Step 4: Translation

In translation, the mRNA message is "read" (also known as translated) & used to build a physical molecule. We must first study ribosomes and the transfer RNA (tRNA) molecule to comprehend translation.

The RNA is *ribosomal RNA or rRNA*, and several proteins comprise a ribosome.

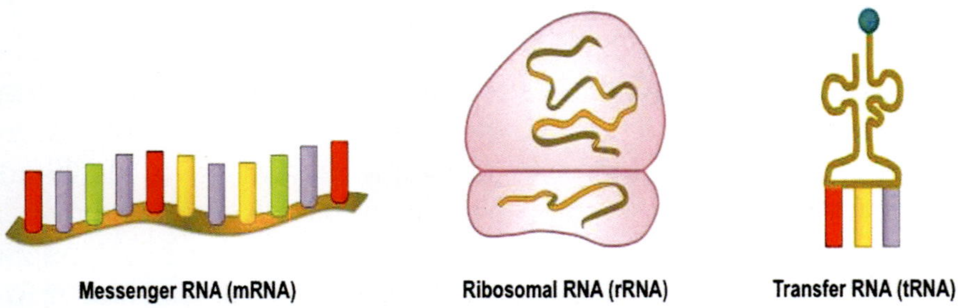

Figure 6.8. Illustration of ribosomal RNA (Source: Bioninja, Creative Commons License).

There is a striking resemblance between the ribosome's physical makeup and its role as a protein maker. The ribosome features a production line and a sizable storage facility. (Muramatsu et al., 1988)

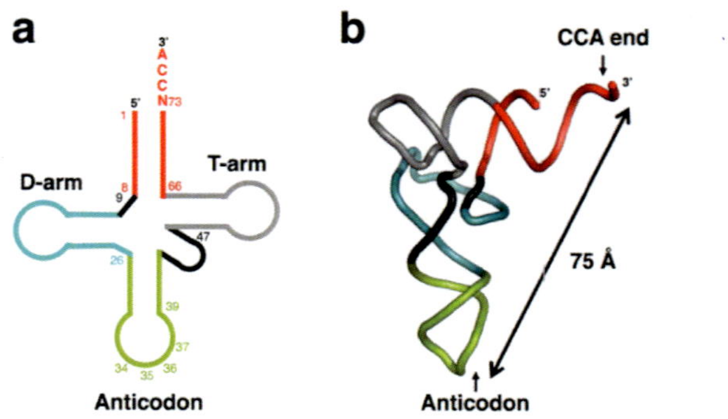

Figure 6.9. Image of transfer RNA: Secondary (a) and tertiary (b) structures of transfer RNA (tRNA). The cloverleaf secondary structure folds to the L-shaped tertiary structure; both structures have corresponding colors. tRNA has the universal single-stranded CCA-3′ terminus and a neighboring nucleotide at position 73 called "discriminator", which is also non-base-paired in most cases. (Source: Tamura K, Creative Commons License).

That contains amino acids, each with a chemical "barcode" which identifies amino acid types connected to it. *Transfer RNA,* often known as *tRNA,* is the RNA-amino acid molecule that contains that chemical barcode (Frye & Blanco, 2016).

A three-nucleotide sequence known as an anticodon serves as a "barcode" that identifies the amino acid that the tRNA molecule carries. The picture refers to the many three-dimensional loops in tRNA as "other RNA." The amino acid itself comes last. Several tRNA molecules are "stored" by each ribosome (Pinho et al., 2009).

Imagine a table with two workers facing each other, one on each side, on the same "factory floor." Around the table, there are several additional employees. An mRNA molecule enters the synthesis facility, binds to the table according to its header information, & moves across it until an mRNA codon known as start codon instructs employees to begin producing a polypeptide chain. One of the workers at the table examines the first mRNA codon before selecting the right tRNA molecule from a bin of tRNA molecules. For instance, the worker will choose a tRNA molecule with anticodon UAC, that transports amino acid methionine, if the initial codon is AUG (coding for Met) (Chen et al., 2020). Next, the mRNA molecule descends the table.

The next mRNA codon as AAA, is read by our worker. The worker binds the complementary tRNA molecule. In this instance, anticodon UUU contains phenylalanine as an amino acid (Phe).

The scientist at the other end of the table connects two amino acids. A peptide bond is used chemically to create the bond (Incarnato, Morandi, Simon, & Oliviero, 2018).

The mRNA is presently traversing the ribosome. The first member of our team performs a scan of the subsequent mRNA codon, which is denoted by the letters AGC, & attaches the appropriate tRNA molecule (one with the anticodon UGR carrying threonine or Thr). The second worker adds the amino acid that was derived from this tRNA to the chain of polypeptides that is now expanding. When this happens, employees working on the production floor separate the amino acid from the tRNA molecule (Kuchino, Borek, Grunberger, Mushinski, & Nishimura, 1982). In the ribosome's storage region, that tRNA will be recycled to pick up a new amino acid & reunite with other tRNA molecules.

This cycle will continue to be repeated until the mRNA codon transforms into a punctuation mark, which will signal the end of the process. The polypeptide chain is now available.

6.5.2.5. Step 5: Post-Translational Modification

It is useful to consider polypeptide chain linear, similar to a train's box cars. A polypeptide's linear structure is significant information, but physics has another plan for our pal. The 20 amino acids that make up proteins have distinct electrical charges and have varying reactions to water, salts, temperature & acidity of the environment where they are synthesized (Krishna & Wold, 1993). Even as the polypeptide chain is leaving the ribosome, these physical pressures compel it to start folding back on itself and taking on a three-dimensional form, sometimes with the help of other molecules. This folding is practically universal and essential for a protein to have an active biological impact. Proteins, such as enzymes and receptors, are able to attach to one another & carry out their function thanks to the three-dimensional, folded structure of the protein, which controls lock-&-key processes that enable this to happen (Deribe, Pawson, & Dikic, 2010).

Polypeptide folding is a common and essential process. However, the polypeptide might undergo additional changes before it develops into a *BAM or biologically active molecule*. Sugar will be added to some proteins (C. Walsh, 2006). They are fat. Sometimes a BAM requires the joining of two or more polypeptide chains. For instance, five polypeptide chains must come together for the nicotinic receptor in the brain to respond to the neurotransmitter acetylcholine (& nicotine, a component in tobacco products). In some circumstances, the freshly translated polypeptide chain must be divided into BAMs (Chou, 2020).

In summary, the polypeptide can become a BAM by various processes that can take place after translation. A post-translational alteration procedure may also be repeated before being undone. Numerous of these systems have protracted and intricate paths. Numerous proteins and enzymes must be turned on for a signaling route to function. As a result, these proteins are constantly activated & then deactivated in response to the cell's signaling requirements (G. Walsh, 2010).

6.6. CASE STUDY: HEMOGLOBIN CHAIN FIRST

This chapter uses hemoglobin protein to demonstrate the genetic code & structure of the genome. It will also benefit us to put the genetic jargon we have learned so far to use (Steinberg, 2008). When we inhale air, our lungs undergo a series of chemical processes that remove oxygen atoms from air & inject them into hemoglobin protein in our red blood cells (Weatherall, 2013).

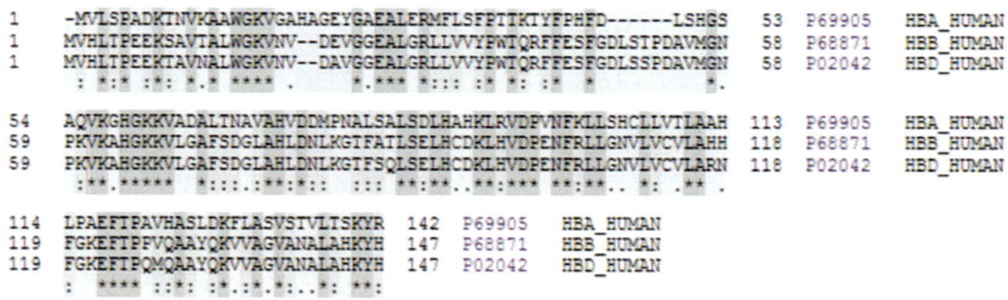

Figure 6.10. Protein alignment of human hemoglobin proteins, alpha, beta, and delta subunits respectively (Source: Wikimedia, Creative Commons License).

When they travel through our arteries, red blood cells finally arrive at microscopic capillaries in bodily tissues like the liver, pancreas, muscle, and neurons, where hemoglobin releases oxygen atoms. Hemoglobin in adults over five months comprises two a chains and two b chains, totaling four polypeptide chains. A different gene codes for each chain (Huisman, Wilson, Gravely, & Hubbard, 1974). Let's investigate the b gene.

A DNA segment carrying gene for b polypeptide is on chromosome 11, this lengthy segment of DNA is over 60,000 nucleotides long (or 60 kb, where kb denotes a *kilobase* or 1,000 base pairs). *Pseudogenes* for the b locus are the boxes marked yb1 and yb2. A similar nucleotide sequence to a functional gene defines a pseudogene whose DNA is not translated or transcribed. A pseudogene doesn't produce a polypeptide that can operate as a biologically active molecule (Dintzis, 1961). Three exons make up the about 1,600 base pair length b coding region. The first exon comprises 90 nucleotides that code for the first 30 amino acids in the peptide chain, the second exon for the next 31 through 104, and the last exon for the final 40 amino acids. Hence, only 438 of 1,600 base pairs carry template information. Because of this, the template and information on how the polypeptide chain should be processed only constitutes around a quarter of the total b locus (Lockard & Lingrel, 1969).

Remember that DNA codons are triplets. The first amino acid in DNAese is valine because the first three coding letters are GTG (Val) (Tran, Silva, & Petrovsky, 2004).

The region of the DNA for the a chains is situated on chromosome 16, which is completely different from the b cluster, and is about 30 kilobases long. The peptide chain template is located in boxes designated a1 and a2. The DNA for both loci is transcribed, altered, & translated into same chains, demonstrating a gene duplication (Waks, Alfsen, Schwaiger, & Mayer, 1969). Similar to the b chain, the a locus also has a pseudogene, which is indicated in Figure 3.8 by the box marked ya1. (Again, disregard boxes marked z1 & z2). Since two a loci have three exons, their true structure is extremely similar to that of the b locus.

Both a genes must be translated into a polypeptide, fold back into a three-dimensional structure, and be edited into mRNA before being translated into a biologically active hemoglobin molecule. Similar processes will result in a folded b polypeptide from the b gene. The joining of two a chains and two b chains (an example of post-translational modification). This molecule has heme groups attached to it (another case of post-translational modification). A unique iron ion found in heme groups can " catch " and attach an oxygen atom to it (Lu, Xu, Zheng, Zhang, & Su, 2006).

Let's engage in a mental exercise to comprehend better how the human genome is structured. Imagine you were enrolled in a biochemical engineering course. Your professor assigns you the task of creating a similar process that can make molecules resembling hemoglobin. What grade would you get if you submitted a sample of human hemoglobin?

Consider this for a second. You would have a sizable template, which would only be partially usable. Moreover, certain areas that could be useful won't be used at all (pseudogenes).

KEYWORD

Hemoglobin is a two-way respiratory carrier, transporting oxygen from the lungs to the tissues and facilitating the return transport of carbon dioxide. In the arterial circulation, hemoglobin has a high affinity for oxygen and a low affinity for carbon dioxide, organic phosphates, and hydrogen and chloride ions.

A non-functional product would result from synthesis anything from a useful part, such as two a sections & the b section. Instead, you must replicate these pieces, clip off the components that render the product useless, and then adhere to the remaining portions (the editing process after transcription).

Also, you struggle with effective output. You will create two usable a "widgets" for every functional b widget if there are only one usable b section and two usable a sections. As a result, you will need additional complex procedures to accelerate the creation of b units while decreasing the production of a units.

The point is that, from an engineering perspective, the human genome is not logical or effective. It seems so erratic that one is baffled about how even a single cell could work, much less develop a functional huge multicellular life form.

KEYWORD

Cell biology is the study of cell structure and function, and it revolves around the concept that the cell is the fundamental unit of life.

6.7. FACTS ABOUT THE HUMAN GENOME

New technological advancements in the 1980s drastically altered genetics and cell biology. During the course of that decade, "holy grail" of human genetics, also known as the discovery of the nucleotide sequence or the order in which the As, Cs, Gs, and Ts are arranged throughout the whole human genome, began to capture attention of geneticists. At the time, it took a long time and money to determine the sequence of even a little portion of DNA. Given the available resources for biological study, it would be unfeasible to attempt this for the entire genome (Gannett, 2001).

Undaunted, groups of geneticists conceived "big science" initiative for medicine & biology comparable to physicists' gigantic particle accelerators & astronomers' space observatories. They approached US Congress, & in 1987, the Department of Energy's human & environmental research projects received beginning funding (DOE). The DOE and the US National Institutes of Health joined forces in 1990 to coordinate this endeavor, & initiative was fully financed for $3 billion over 15 years. Soon after, institutions from the United Kingdom, Japan, & other developed countries joined the Human Genome Project (HGP) (Read, 2009).

A rough sequence form was announced in 2000 at a news conference led by US President Bill Clinton and British Prime Minister Tony Blair due to developments in biology and computing science. The complete sequence was published three years later. Unlike many government-funded projects, the HGP was completed on time and within budget. The following are some of the most important findings from the HGP (Watson, 2001).

A consensus sequence. A single human genome sequence does not exist. There're as many sequences of the human genome as there have been members of the Homo sapiens species, with the exception of identical twins. Instead, the HGP synthesized a consensus sequence that serves as a map against which geneticists can compare other genomes. Also, it is a composite sequence synthesized from the genomes of several individuals (McCann-Mortimer, Augoustinos, & LeCouteur, 2004).

3.2 billion nucleotides. The human genome sequence contains 3.2 billion nucleotides of information. Look at just one cell in a male human to see what this entails. The cell has two copies of the autosomal chromosomes. We now have 21 autosomal chromosomes, X & Y chromosomes, and one copy of each copy of each gene. They are all double-stranded. Let's remove the second strand and make them single-stranded. There would be 3.2 billion nucleotides left over (Choi, Chan, Kirkness, Telenti, & Schork, 2018).

KEYWORD

Human genome is a complete set of nucleic acid sequences for humans, encoded as DNA within the 23 chromosome pairs in cell nuclei and in a small DNA molecule found within individual mitochondria.

20,000 to 25,000 genes. A gene is a section of DNA that, according to the definitions of molecular biologists, contains the instructions for the formation of a polypeptide chain. According to this definition, human beings contain somewhere between 20,000 & 25,000 genes, which is a significantly lower number than what was anticipated when the study first started (Caplan, 1998).

Roughly 2% of the human genome contains the template for polypeptides. It is simple to determine a rough estimate of the percentage of the human genome that codes for proteins, given the number of genes & average size of a gene. That turns out to be a relatively modest percentage—about 2%. The fact that template content makes up a very small portion of our total DNA does not change much, even after including the introns, promoter regions, DNA that codes for different RNAs (Di Minno & de Gaetano, 2013).

Most human DNA has no discernible function. What is the function of the remaining DNA if just 2% of it codes for proteins? The unexpected response is that we don't know. However, there are undoubtedly some locations that serve as regulatory areas. In other words, they adjust the protein-coding areas' "dimmer switch." But, even if we discovered twice or three times as many of these previously unknown regions, we would still fall short of fully understanding the remaining 98% of human DNA (Yamagishi &

Shimabukuro, 2008). *An unknown amount of our DNA is probably junk.* A DNA segment may have a purpose even if we cannot determine what that purpose is. Hence, it isn't easy to estimate the proportion of the human genome which is not functional. The genome still has substantial repeating regions that can be removed or inverted without appearing to have any effect. There are also about 10,000 pseudogenes. There is widespread agreement among geneticists that certain regions of human DNA are junk (i.e., non-functional), although there's disagreement on this proportion (Cunningham, 1998).

ACTIVITY 6.1.

A forensic scientist is trying to identify the perpetrator of a crime by analyzing the DNA evidence found at the crime scene. What are the different methods used to analyze DNA, and what are the limitations of these techniques?

SUMMARY

Chapter "DNA and RNA" covers various aspects related to the structure, function, and significance of DNA and RNA in the development of human beings. The learning objectives of this chapter include the physical structure of DNA, the importance of DNA for human development, the concept of RNA and genetic code, the process of protein synthesis, and a case study on hemoglobin chains.

The chapter begins by introducing the physical structure of DNA, including the double helix structure and the role of base pairing in DNA replication. It then discusses the importance of DNA in the development of human beings, including its role in determining traits and susceptibility to diseases.

Next, the chapter covers the concept of RNA and the genetic code, including the different types of RNA and the codons that code for specific amino acids. The process of protein synthesis is also discussed, including the roles of transcription and translation.

Finally, the chapter presents a case study on hemoglobin chains, which illustrates the importance of DNA and RNA in the development of complex organisms. It explains how changes in the genetic code can result in genetic disorders such as sickle cell anemia. Overall, this chapter provides a comprehensive overview of the structure, function, and significance of DNA and RNA in human biology.

REVIEW QUESTIONS

1. What is the physical structure of DNA and how is it organized?
2. What is the importance of DNA in the development of humans and how does it relate to genetic diseases?
3. What is RNA and how does it differ from DNA? How is the genetic code translated from DNA to RNA?
4. What is the process of protein synthesis and how is it regulated? What role do ribosomes play in this process?
5. How does the case study of the hemoglobin chain illustrate the relationship between DNA, RNA, and protein synthesis? What are some genetic mutations that can affect hemoglobin production and cause disease?

MULTIPLE CHOICE QUESTIONS

1. What is the physical structure of DNA?

a. Single-stranded helix

b. Double-stranded helix

c. Triple-stranded helix

d. Quadruple-stranded helix

2. What is the importance of DNA for the development of humans?

a. It controls the expression of genes.

b. It stores genetic information.

c. It replicates itself during cell division.

d. All of the above.

3. What is RNA?

a. A single-stranded molecule that stores genetic information

b. A double-stranded molecule that stores genetic information

c. A single-stranded molecule that transfers genetic information from DNA to the ribosome

d. A double-stranded molecule that transfers genetic information from DNA to the ribosome

4. What is the genetic code?

a. The set of rules that determine the sequence of nucleotides in DNA

b. The set of rules that determine the sequence of amino acids in proteins

c. The set of rules that determine the structure of RNA

d. The set of rules that determine the shape of DNA

5. What is the process of protein synthesis?

a. Transcription followed by translation

b. Translation followed by transcription

c. Replication followed by translation

d. Translation followed by replication

6. What is the case study of the hemoglobin chain?

a. It demonstrates the importance of DNA in controlling gene expression.

b. It illustrates the process of protein synthesis.

c. It shows the relationship between DNA and RNA.

d. It highlights the importance of mutations in genetic diseases.

Answers to Multiple Choice Questions

1. (b) 2. (d) 3. (c) 4. (b) 5. (a)

6. (b)

1.

REFERENCES

1. Bucholtz, M. (2000). The politics of transcription. *Journal of pragmatics, 32*(10), 1439-1465.
2. Burgers, P. M. (1998). Eukaryotic DNA polymerases in DNA replication and DNA repair. *Chromosoma, 107*(4), 218-227.
3. Caplan, A. L. (1998). What's so special about the human genome? *Cambridge Quarterly of Healthcare Ethics, 7*(4), 422-424.
4. Carter, C., & Houlihan, D. (2001). Protein synthesis. *Fish Physiology, 20*, 31-75.
5. Chen, Z., Zhao, P., Li, F., Wang, Y., Smith, A. I., Webb, G. I., . . . Song, J. (2020). Comprehensive review and assessment of computational methods for predicting RNA post-transcriptional modification sites from RNA sequences. *Briefings in Bioinformatics, 21*(5), 1676-1696.
6. Choi, Y., Chan, A. P., Kirkness, E., Telenti, A., & Schork, N. J. (2018). Comparison of phasing strategies for whole human genomes. *PLoS Genetics, 14*(4), e1007308.
7. Choo, Y., & Klug, A. (1997). Physical basis of a protein-DNA recognition code. *Current Opinion in structural biology, 7*(1), 117-125.
8. Chou, K.-C. (2020). Progresses in predicting post-translational modification. *International Journal of Peptide Research and Therapeutics, 26*(2), 873-888.
9. Cleaver, J. (1968). Defective repair replication of DNA in xeroderma pigmentosum. *Nature, 218*(5142), 652-656.
10. Collie, G. W., & Parkinson, G. N. (2011). The application of DNA and RNA G-quadruplexes to therapeutic medicines. *Chemical Society Reviews, 40*(12), 5867-5892.
11. Crick, F. H. (1958). *On protein synthesis.* Paper presented at the Symp Soc Exp Biol.
12. Crick, F. H. (1966). *The genetic code—yesterday, today, and tomorrow.* Paper presented at the Cold Spring Harbor symposia on quantitative biology.
13. Crick, F. H. (1968). The origin of the genetic code. *Journal of Molecular Biology, 38*(3), 367-379.
14. Crick, F. H., Barnett, L., Brenner, S., & Watts-Tobin, R. J. (1961). General nature of the genetic code for proteins. *Nature, 192*, 1227-1232.
15. Cunningham, H. (1998). Colonial encounters in postcolonial contexts: Patenting indigenous DNA and the Human Genome Diversity Project. *Critique of Anthropology, 18*(2), 205-233.
16. Deribe, Y. L., Pawson, T., & Dikic, I. (2010). Post-translational modifications in signal integration. *Nature Structural & Molecular Biology, 17*(6), 666-672.
17. Di Minno, G., & de Gaetano, G. (2013). Human genome, environment and medical practice. *Internal and Emergency Medicine, 8*, 645-649.
18. Dintzis, H. M. (1961). Assembly of the peptide chains of hemoglobin. *Proceedings of the National Academy of Sciences, 47*(3), 247-261.

19. Freeland, S. J., & Hurst, L. D. (1998). The genetic code is one in a million. *Journal of Molecular Evolution, 47*(3), 238.

20. Frye, M., & Blanco, S. (2016). Post-transcriptional modifications in development and stem cells. *Development, 143*(21), 3871-3881.

21. Gannett, L. (2001). Racism and human genome diversity research: the ethical limits of "population thinking". *Philosophy of Science, 68*(S3), S479-S492.

22. Garner, M. H., Bogardt Jr, R., & Gurd, F. (1975). Determination of the pK values for the alpha-amino groups of human hemoglobin. *Journal of Biological Chemistry, 250*(12), 4398-4404.

23. Geider, K., & Hoffmann-Berling, H. (1981). Proteins controlling the helical structure of DNA. *Annual Review of Biochemistry, 50*(1), 233-260.

24. Hager, G. L., McNally, J. G., & Misteli, T. (2009). Transcription dynamics. *Molecular Cell, 35*(6), 741-753.

25. Harland, R. M., & Laskey, R. A. (1980). Regulated replication of DNA microinjected into eggs of *Xenopus laevis*. *Cell, 21*(3), 761-771.

26. Haselkorn, R., & Rothman-Denes, L. (1973). Protein synthesis. *Annual Review of Biochemistry, 42*(1), 397-438.

27. Houlihan, D., Carter, C., & McCarthy, I. (1995). Protein synthesis in fish *Biochemistry and Molecular Biology of Fishes* (Vol. 4, pp. 191-220): Elsevier.

28. Huisman, T., Wilson, J., Gravely, M., & Hubbard, M. (1974). Hemoglobin Grady: the first example of a variant with elongated chains due to an insertion of residues. *Proceedings of the National Academy of Sciences, 71*(8), 3270-3273.

29. Hurwitz, J., Dean, F., Kwong, A., & Lee, S. (1990). The *in vitro* replication of DNA containing the SV40 origin. *Journal of Biological Chemistry, 265*(30), 18043-18046.

30. Incarnato, D., Morandi, E., Simon, L. M., & Oliviero, S. (2018). RNA Framework: an all-in-one toolkit for the analysis of RNA structures and post-transcriptional modifications. *Nucleic Acids Research, 46*(16), e97-e97.

31. Joachimi, A., Benz, A., & Hartig, J. S. (2009). A comparison of DNA and RNA quadruplex structures and stabilities. *Bioorganic & Medicinal Chemistry, 17*(19), 6811-6815.

32. Kiss, T. (2001). Small nucleolar RNA-guided post-transcriptional modification of cellular RNAs. *The EMBO Journal, 20*(14), 3617-3622.

33. Krishna, R. G., & Wold, F. (1993). Post-translational modifications of proteins. *Methods in Protein Sequence Analysis*, 167-172.

34. Kuchino, Y., Borek, E., Grunberger, D., Mushinski, J. F., & Nishimura, S. (1982). Changes of post-transcriptional modification of wye base in tumor-specific tRNAPhe. *Nucleic Acids Research, 10*(20), 6421-6432.

35. Latchman, D. S. (1997). Transcription factors: an overview. *The International Journal of Biochemistry & Cell Biology, 29*(12), 1305-1312.

36. Lockard, R. E., & Lingrel, J. B. (1969). The synthesis of mouse hemoglobin chains in a rabbit reticulocyte cell-free system programmed with mouse reticulocyte 9S RNA. *Biochemical and Biophysical Research Communications, 37*(2), 204-212.
37. López-Contreras, A. J., & Fernandez-Capetillo, O. (2010). The ATR barrier to replication-born DNA damage. *DNA Repair, 9*(12), 1249-1255.
38. Lu, X., Xu, Y., Zheng, C., Zhang, G., & Su, Z. (2006). Ethylene glycol diglycidyl ether as a protein cross-linker: a case study for cross-linking of hemoglobin. *Journal of Chemical Technology & Biotechnology: International Research in Process, Environmental & Clean Technology, 81*(5), 767-775.
39. Machnicka, M. A., Olchowik, A., Grosjean, H., & Bujnicki, J. M. (2014). Distribution and frequencies of post-transcriptional modifications in tRNAs. *RNA Biology, 11*(12), 1619-1629.
40. McCann-Mortimer, P., Augoustinos, M., & LeCouteur, A. (2004). 'Race' and the Human Genome Project: constructions of scientific legitimacy. *Discourse & Society, 15*(4), 409-432.
41. Meselson, M., & Stahl, F. W. (1958). *The replication of DNA.* Paper presented at the Cold Spring Harbor symposia on quantitative biology.
42. Meselson, M., & Stahl, F. W. (1958). The replication of DNA in *Escherichia coli. Proceedings of the National Academy of Sciences, 44*(7), 671-682.
43. Moldave, K. (1985). Eukaryotic protein synthesis. *Annual Review of Biochemistry, 54*(1), 1109-1149.
44. Muramatsu, T., Nishikawa, K., Nemoto, F., Kuchino, Y., Nishimura, S., Miyazawa, T., & Yokoyama, S. (1988). Codon and amino-acid specificities of a transfer RNA are both converted by a single post-transcriptional modification. *Nature, 336*(6195), 179-181.
45. Nachtergaele, S., & He, C. (2017). The emerging biology of RNA post-transcriptional modifications. *RNA Biology, 14*(2), 156-163.
46. Nirenberg, M. W. (1963). The genetic code. *Scientific American, 208*(3), 80-95.
47. Oerum, S., Dégut, C., Barraud, P., & Tisné, C. (2017). m1A post-transcriptional modification in tRNAs. *Biomolecules, 7*(1), 20.
48. Palade, G. (1975). Intracellular aspects of the process of protein synthesis. *Science, 189*(4200), 347-358.
49. Pinho, S. S., Reis, C. A., Paredes, J., Magalhaes, A. M., Ferreira, A. C., Figueiredo, J., . . . Seruca, R. (2009). The role of N-acetylglucosaminyltransferase III and V in the post-transcriptional modifications of E-cadherin. *Human Molecular Genetics, 18*(14), 2599-2608.
50. Proud, C. (2006). Regulation of protein synthesis by insulin. *Biochemical Society Transactions, 34*(2), 213-216.

51. Räschle, M., Knipscheer, P., Enoiu, M., Angelov, T., Sun, J., Griffith, J. D., . . . Walter, J. C. (2008). Mechanism of replication-coupled DNA interstrand crosslink repair. *Cell, 134*(6), 969-980.

52. Read, A. P. (2009). Stewart Scherer: A short guide to the human genome: Cold Spring Harbor Laboratory Press (2008), ISBN 978-087969791-4, $29.00, 173 pp: Springer.

53. Steinberg, M. H. (2008). Sickle cell anemia, the first molecular disease: overview of molecular etiology, pathophysiology, and therapeutic approaches. *The Scientific World Journal, 8*, 1295-1324.

54. Tran, H. A., Silva, D., & Petrovsky, N. (2004). Case study: potential pitfalls of using hemoglobin A1c as the sole measure of glycemic control. *Clinical Diabetes, 22*(3), 141-143.

55. Waks, M., Alfsen, A., Schwaiger, S., & Mayer, A. (1969). Structural studies of haptoglobins: III. Interactions with hemoglobin. *Archives of Biochemistry and Biophysics, 132*(1), 268-278.

56. Walsh, C. (2006). *Posttranslational modification of proteins: expanding nature's inventory*: Roberts and Company Publishers. ISBN: 0974707732, 9780974707730.

57. Walsh, G. (2010). Post-translational modifications of protein biopharmaceuticals. *Drug discovery today, 15*(17-18), 773-780.

58. Wang, L., Brock, A., Herberich, B., & Schultz, P. G. (2001). Expanding the genetic code of *Escherichia coli. Science, 292*(5516), 498-500.

59. Wang, L., & Schultz, P. G. (2002). Expanding the genetic code. *Chemical Communications*(1), 1-11.

60. Wang, L., & Schultz, P. G. (2005). Expanding the genetic code. *Angewandte Chemie International Edition, 44*(1), 34-66.

61. Wang, L., Xie, J., & Schultz, P. G. (2006). Expanding the genetic code. *Annu. Rev. Biophys. Biomol. Struct., 35*, 225-249.

62. Watson, J. D. (2001). The human genome revealed. *Genome Research, 11*(11), 1803-1804.

63. Watson, J. D., & Crick, F. H. (1953). *The structure of DNA.* Paper presented at the Cold Spring Harbor symposia on quantitative biology.

64. Weatherall, D. J. (2013). The role of the inherited disorders of hemoglobin, the first "molecular diseases," in the future of human genetics. *Annual Review of Genomics and Human Genetics, 14*, 1-24.

65. Wigler, M., Levy, D., & Perucho, M. (1981). The somatic replication of DNA methylation. *Cell, 24*(1), 33-40.

66. Wilken, J., & Kent, S. B. (1998). Chemical protein synthesis. *Current Opinion in Biotechnology, 9*(4), 412-426.

67. Woese, C. R. (1965). On the evolution of the genetic code. *Proceedings of the National Academy of Sciences, 54*(6), 1546-1552.

68. Yamagishi, M. E. B., & Shimabukuro, A. I. (2008). Nucleotide frequencies in human genome and Fibonacci numbers. *Bulletin of Mathematical Biology, 70*(3), 643-653.

69. Zimmerman, S. B. (1982). The three-dimensional structure of DNA. *Annual Review of Biochemistry, 51*(1), 395-427.

70. Zwolak, M., & Di Ventra, M. (2008). Colloquium: Physical approaches to DNA sequencing and detection. *Reviews of Modern Physics, 80*(1), 141.

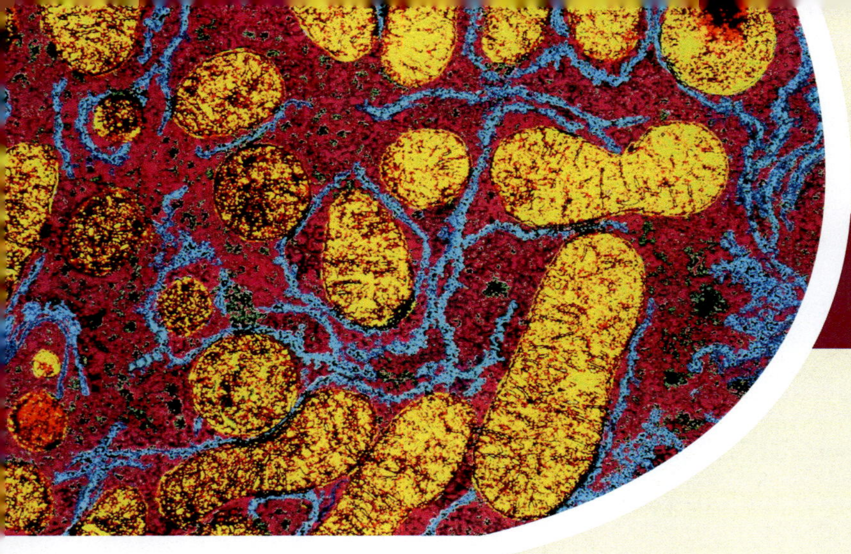

CHAPTER 7

METABOLIC PATHWAYS IN THE HUMAN BODY

INTRODUCTION

The primary metabolic pathways for various biological components, including energy & carbohydrates metabolism via electron transfer systems, lipoproteins, amino acids, lipids & creation of nucleic acids & proteins, are discussed in this chapter. A disruption in the metabolic pathways or control of those pathways is the root cause of the metabolic syndrome. Diabetes mellitus is a condition that can be brought on by disturbances in the anaerobic metabolism of sugar as well as the glycogen metabolism. Diseases such as Parkinson's & Alzheimer's are problems that might occur with electron transport systems. Regarding the composition of sugar chains, some blood group chemicals can be utilized as helpful markers for tumors (Wu et al., 2010). A disorder called paroxysmal nocturnal hemoglobinuria, which is caused by a dysfunction in the ability of glycosylphosphatidylinositol group to connect to asparagine-linked sugar chains, can be a potentially fatal illness. Both cholesterol and lipid metabolism affect a person's presence or absence of disease. The metabolic processing of lipids is an important marker of lipoprotein & fatty acids. Disorders in the metabolism of nucleic acids & amino acids and diseases that result from these disorders are illustrated here. A synopsis of the protein production process is shown here (Rae, 2014).

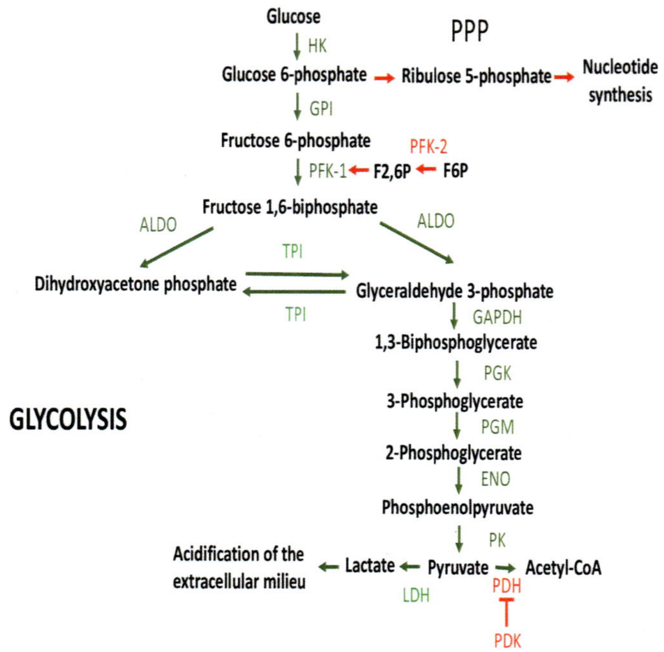

Figure 7.1. Glycolysis. Glycolytic enzymes and metabolites are depicted as well as some of the pathways that branch off or that modulate steps in the glycolytic metabolism (see text for more details). Green arrows refer to steps that are inherent to glycolysis, red arrows to reactions that branch off or modulate glycolysis. ALDO, aldolase, fructose-bisphosphate; ENO, enolase; GAPDH, glyceraldehyde 3-phosphate dehydrogenase; GPI, glucose 6-phosphate isomerase; HK, hexokinase; LDH, lactate dehydrogenase; PDH, pyruvate dehydrogenase; PDK, pyruvate dehydrogenase kinase; PFK-1, phosphofructokinase-1; PFK-2, phosphofructokinase-2 (including the isoforms 6-phosphofructo-2-kinase/fructose-2,6-biphosphatase 3 (PFKFB3) and PFKFB4); PGK, phosphoglycerate kinase; PGM, phosphoglycerate mutase; PK, pyruvate kinase; PPP, pentose phosphate pathway; TPI, triosephosphate isomerase. (Source: Marcucci F & Rumio C, Creative Commons License).

Learning Objectives

At the end of this chapter, the reader will be able to understand

- The basics of sugar metabolism, including glycolysis, the TCA cycle, and glycogen metabolism.
- The process of gluconeogenesis and its importance in maintaining blood glucose levels.
- The role of the pentose phosphate pathway in generating NADPH and ribose-5-phosphate.
- The blood groups and understand their importance in blood transfusions.
- The basics of lipid metabolism, including the synthesis and breakdown of lipids and the role of lipoproteins.

- The synthesis and functions of steroid hormones.
- The importance of amino acid metabolism, including the roles of leucine, isoleucine, valine, phenylalanine, and tyrosine.

Key Terms

- Glycolysis
- TCA Cycle
- Glycogen
- Gluconeogenesis
- Pentose Phosphate Pathway
- Blood Groups
- Lipoprotein
- Steroid Hormone
- Amino Acid Metabolism
- Urea Cycle
- Purine Metabolism
- Protein Biosynthesis

7.1. SUGAR METABOLISM

Figure 7.2 depicts the pathway of sugar absorption in the small intestine.

7.1.1. Glycolysis

7.1.1.1. Preparatory Stage

The first 5 steps of glycolysis are known as energy or preparation investment phase. Adenosine triphosphate (ATP) (two molecules) are provided during the preparation phase, and glucose is phosphorylated & converted to glyceraldehyde-3-phosphate (G3P) (Muhammad et al., 2013).

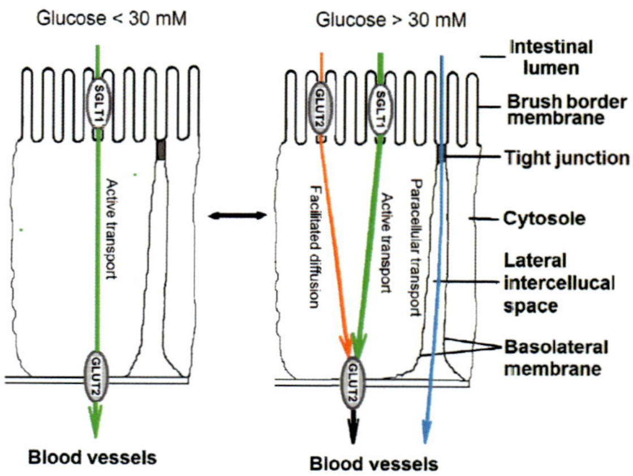

Figure 7.2. Schematic of the mechanism of absorption of sugar in the small intestine (Source: Gromova et al., Creative Commons License).

Initially in the preparatory phase, an enzyme known as hexokinase is responsible for phosphorylating glucose. The glucose-6-phosphate formation is the end product of an ATP reaction in which a carbon 6 (C6) position is moved from glucose to g-phosphate residue (G6P). The magnesium ion (Mg2+) is necessary for this process. Hexokinase comes in various forms (isozymes). Even though the catalyst speeds up the same reaction, isozymes are encoded by multiple genes. Hexakinase exists in four isozymes in mammals. Varying isozymes have a varying affinity for glucose (different Michaelis constants or Km values). KM value of Hexokinase I, II, &III is 106 M. Km value of glucokinase, also known as hexokinase IV, has generally 102 M, but this value might vary depending on how glucokinase works. Hepatic cells contain a high quantity of glucokinase (Dipietro et al., 1962).

Glucokinase does not generally completely limit glucose production since the concentration of glucose in the blood is normally lower than the Km. In this case, different isozymes of hexokinase are responsible for catalyzing the process. Glucokinase starts doing its job when there is a higher concentration of glucose in the blood. Because glucokinase is rarely saturated, glucose can travel swiftly to the glycolytic pathway or glycogen synthesis shunt even when the quantity of glucose in hepatic cells rises dramatically. The amount of glucose that is contained within the cell is typically not very high, and the presence of extracellular glucose prevents glucose from leaking out of the cell into the surrounding environment while simultaneously facilitating membrane transfer into the cell. Insulin has an ability to stimulate the activity of the enzyme hexokinase. In contrast, glucagon can inhibit hexokinase activity, causing glucose-6-phosphatase activity to increase. *In vivo*, gluconeogenesis will be stimulated (Bonadonna et al., 1996).

Remember

From the sweetness of indulgence to the bitterness of metabolic imbalance, sugar's journey through our body leaves a lasting impression on our health and well-being.

Stage 2: Glucose-6-Phosphate Isomerization

Phosphoglucoisomerase (phosphoglucose isomerase) transforms glucose-6-phosphate (G6P) to fructose-6-phosphate (F6P) in 2nd stage. Mg2+ is also required for this reaction. From the following stage, F6P can enter the glycolytic pathway. The free energy charge in this reaction is modest. It can proceed in either direction, but the opposite reaction is rare because F6P is rapidly consumed in the following step. a-Dglucopyranose-6-phosphate, which is the a-anomer of G6P, preferentially interacts with enzyme. A ring, therefore, opens, converting an aldose into ketose. After this stage, the isomerization reaction that follows is an essential part of the glycolytic pathway (Hoffmann et al., 2014).

Stage 3: Fructose-6-Phosphate Phosphorylation

In the third stage, fructose-1,6-bisphosphate (F1,6BP) is produced when phosphate from ATP is transferred to the C1 hydroxyl residue of fructose-6-phosphate by an enzyme known as phosphofructokinase-1 (PFK-1). Mg2+ & fructose-2,6-bisphosphate are required for this process. Fructose-2,6-bisphosphate levels increase after a meal, causing the previously mentioned reaction to promote fast glycolysis to occur (Sakakibara et al., 1997). Fructose-2,6-bisphosphate concentration is reduced before a meal, leading to the facilitation of gluconeogenesis & upregulation of F1,6BP. In contrast to F6P and G6P, which can be broken down by metabolic routes other than the glycolytic system, F1,6BP can only be broken down via the glycolytic system. It is impossible to undo the reaction since the reaction where PFK-1 works as a catalyst is irreversible, &

PFK-1 doesn't utilize a reverse reaction in the glyconeogenesis process. As a result, two enzyme processes regulated by fructose-2,6-bisphosphate are important components of glyconeogenesis & glycolytic systems (Pilkis et al., 1982).

Stage 4: Cleavage

1st three glycolysis reactions are called preliminary processes because they cleave F1,6BP & produce 2rd triose phosphates. Those are the products of glycolysis. F1,6BP is cleaved into G3P & dihydroxyacetone phosphate (DHAP) in this stage by F1,6BP aldolase (or simply aldolase). Because aldolase is the enzyme that catalyzes this activity, the standard free energy of the cell undergoes a significant and favorable change in the direction of F1,6BP cleavage as a consequence (Enyedi, 2005). The reaction is primarily in equilibrium, & it doesn't serve as the glycolysis control point. When there's a low concentration of intracellular metabolites, there is a little change in free energy, & reverse reaction occurs rather fast. Type I aldolase is found in plants & animals, while type II aldolase is found in bacteria & fungi. A cleavage mechanism of hexose differs between the two forms (Ford & Jennings, 2020).

KEYWORD

Dihydroxyacetone phosphate is the anion with the formula $HOCH_2CCH_2OPO_3^{2-}$. This anion is involved in many metabolic pathways, including the Calvin cycle in plants and glycolysis. It is the phosphate ester of dihydroxyacetone.

Stage 5: Triose Phosphate Isomerization

Out of the two molecules mentioned in the paragraph above, only G3P acts as a substrate in future steps of the process. On the other hand, triose phosphate isomerase catalyzes a reversible mechanism that swiftly transforms DHAP into G3P. Due to the role that triose phosphate isomerase plays as a catalyst in a specific solid target process, the only isomer that is produced is D-the isomer. G3P's carbon 1 (C1) position is obtained from C3 of DHAP, while G3P's C3 & C2 positions are generated from glucose's C2 and C1 positions. Following step 4, glyceraldehyde-3-phosphate is formed, and its C1, C2, & C3 locations correspond to C4, C5, & C6 positions of glucose, respectively. However, separation doesn't systematically separate carbon at each position of 2 G3Ps in a chemical sense. This reaction produces two molecules of G3P from the hexose molecule, & glycolysis preparatory phase is completed (Richard, 1984).

7.1.1.2. Payoff Phase

The energy or payoff phase comprises the five reactions that include 2nd half of glycolysis. During a phase known as the payoff phase, two G3P molecules are converted into pyruvate, 4 molecules of adenosine diphosphate (ADP) are converted into 4 molecules of ATP, & a portion of free energy that was previously contained in glucose is preserved. Since 2 molecules of ATP were used up in the initial phase, glycolysis will result in a net gain of ATP's two molecules. Additionally, each glucose molecule generates 2 nicotinamide adenine dinucleotide (NADH) molecules (Hinckelmann et al., 2016).

KEYWORD

Adenosine diphosphate (ADP) is a nucleoside phosphate comprised of a ribonucleoside and two phosphate groups. It means it has a ribose as its sugar and two phosphate groups attached. Its nucleoside contains a purine base, i.e. an adenine attached to the ribose sugar.

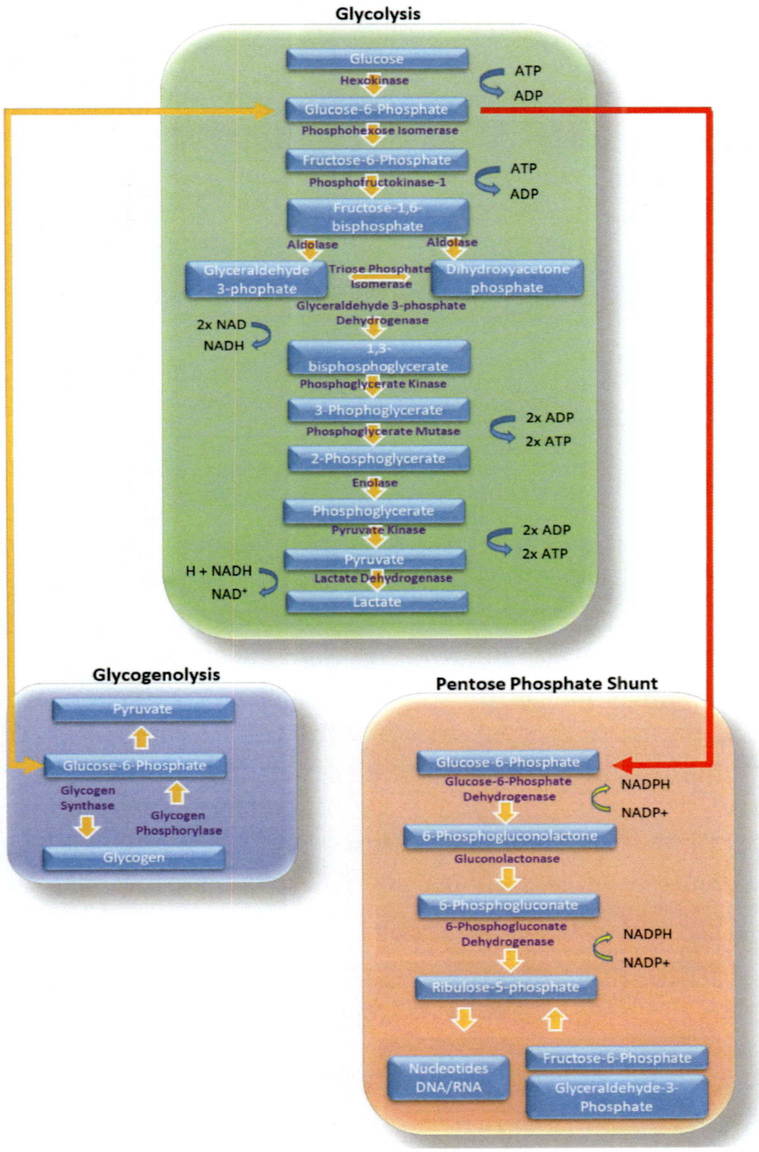

Figure 7.3. Diagram of glycolysis (Source: Bell et al, Creative Commons License).

Stage 6: Glyceraldehyde-3-Phosphate Oxidation

Glyceraldehyde3-phosphate dehydrogenase catalyzes conversion of G3P into 1,3-bisphosphoglycerate in step at start of payoff phase.

Glyceraldehyde-3-phosphoric acid (NADH) undergoes dehydration of its aldehyde residue, and one molecule of nicotinamide adenine dinucleotide (NAD+), which is in an oxidized state, is transformed into a reduced form of NADH. The aldehyde group oxidation reduces standard free energy in G3P, & a large portion of this reduced energy will be stored in an acyl phosphate. The energy released during the hydrolysis of acyl phosphate, which has an ester link between carboxylic acid and an acyl residue (R-CO-), is particularly quick (Arutyunova et al., 2003). The next stage, which requires this energy, produces ATP from ADP. When heavy metals, like mercury (Hg2+), combine with cysteine found in an enzyme's active site, the process that the enzyme is responsible for is irreversibly blocked. Renal failure is induced by Hg2+ toxicity. A process will end if NADH oxidizes once more & NAD+ isn't continuously supplied because the NAD+ amount in the cell is limited. Lactate fermentation & alcohol are two examples of processes that can lead to the production of NADH by reoxidation (Muronetz et al., 2020).

KEYWORD

Phosphate residue is a kind of hazardous solid waste and if not properly disposed of, could cause serious environmental contaminations. The abundant iron salt available in phosphate residue can be used to prepare photo-Fenton catalytic reagent for wastewater treatment.

Stage 7: Phosphate Residue Transfer from 1,3-Bisphosphoglycerate to Adenosine Diphosphate

A process based on phosphoglycerate kinase (PGK) makes up 2nd stage of the payoff phase. Phosphate residue, which contains greater energy, travels from 1,3-bisphosphoglycerate to ADP during the process that is catalyzed by PGK and results in the production of ATP & 3-phosphoglycerate. This is the phosphorylation that occurs at the substrate level during glycolysis. Mg2+ is necessary for the reaction. The energy coupling reaction consists of aldehyde oxidation in stage 6, and ATP synthesis in stage 7, & it uses 1,3-bisphosphoglyceric acid as an intermediate (Blake & Rice, 1981).The first acyl phosphate formation reaction is endergonic, & subsequent ATP production is a powerful exergonic reaction. The following formula can be used to express the energy when the two reactions are added together:

Glyceraldehyde-3-phosphate + ADP + +Pi NAD+ 3-Phosphoglycerate + ATP + NADH+ H^+

Pi stands for inorganic phosphate. It is generally an exergonic response (Ayna & Moody, 2021).

Stage 8: Intramolecular Rearrangement of the Phosphate Residue

Well, this stage involves intramolecular rearrangement reaction catalyzed by mutase rather than isomerization. Using the enzyme catalyst phosphoglycerate mutase (PGM), phosphate residue in glycerate is transferred reversibly and converted to 2-phosphoglycerate. Mg2+ is needed for this reaction. There are two steps in this reaction. Animals and plants have different reaction processes. The intermediate reaction 2,3-bisphosphoglycerate (2,3-BPG or just BPG) is produced when phosphate that is connected to His residue of the active center of the enzyme transfers to the hydroxyl residue of C2 position of the 3-phosphoglycerate, this occurs in animals. A phosphorylated enzyme is then replicated, leading to the generation of reaction product 2-phosphoglycerate, as transferred phosphate residue is transferred back to His residue from C3 of 2,3-BPG. Because to phosphorylate PGM 2,3-BPG is required, the cell must always keep a trace quantity of it in storage so that it may initiate the catalytic activity. This trace amount must be quite modest (Kuzuyama et al., 2000).

In contrast to animals, plants do not create intermediate 2,3-BPG. In a nutshell, 3-phosphoglycerate first attaches to the enzyme, & then phosphate residue moves to the active center. This phosphate residue is then brought back to C2, creating 2-Phosphoglycerate (Hodgson & Petroliagi, 2001).

Stage 9: 2-Phosphoglycerate Dehydration

A high phosphate transition potential molecule is made in the 4th step of the payoff phase. Enolase is responsible for catalyzing the process, which transforms 2-phosphoglycerate into phosphoenolpyruvate by reversibly using water (H_2O) from the C2 and C3 of the compound. (PEP). A phosphate transition potential is extremely high because the enol in pyruvate is unstable. In 3-phosphoglycerate, the amount of standard free energy necessary for phosphate group hydrolysis is $\Delta G = -17.6\ kJ/mol$, but in PEP, the amount of standard free energy required is $\Delta G = -61.9 kJ/mol$. However, total calories in both molecules are nearly equal, dehydration reaction redistributes the energy. Two Mg^{2+} must participate in this reaction. As the "conformational" ion, one Mg^{2+} attaches to the carboxyl residue of the substrate. As a "catalyst" ion, another Mg^{2+} binds to phosphate & carboxyl residues. These two ions prevent oxidation, and the lysine residue in the active center of enolase pulls out the hydrogen atom from the C2 position, the hydrogen atom from the glutamate residue, & the hydroxyl group from the C3 position from H_2O (Wang & Himoe, 1974).

KEYWORD

TCA cycle is the series of chemical reactions used by all aerobic organisms to release stored energy through the oxidation of acetyl CoA derived from carbohydrates, fats, and proteins into ATP." TCA cycle or Tricarboxylic Cycle is also known as Kreb's Cycle or Citric Acid Cycle.

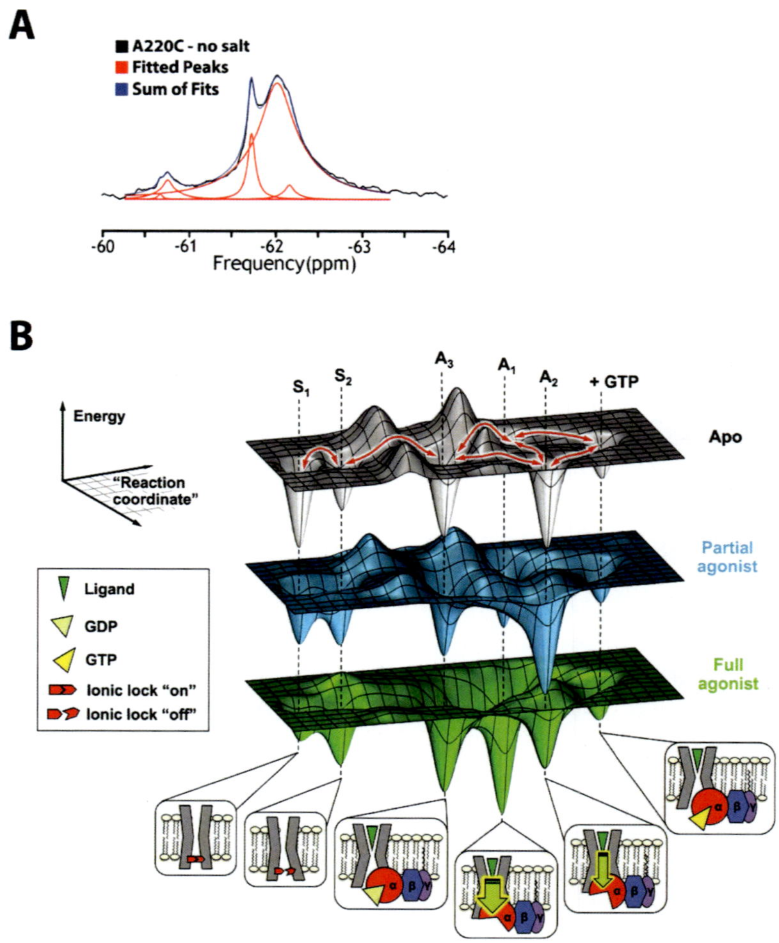

Figure 7.4. Image of the conformational transition of an ion (Source: Puthenveetil et al., Creative Commons License).

Stage 10: Phosphate Residue Transfer from Phosphoenolpyruvate

A transition pyruvate kinase in cells produces the enolic form of pyruvate, which is then swiftly transformed into keto form, an isomer of PEP (Tovar et al., 1997).

The final step is the transfer of a phosphate residue from PEP to ADP by enzyme catalyst pyruvate kinase. Pyruvate & ATP are produced as a result. Either magnesium (Mg^{2+}) & potassium (K^{+}) or manganese (Mn^{2+}) ions are needed for this process. The concentration of the second substrate affects the phosphorylation process. This reaction is irreversible and serves as a crucial regulating point depending on the intracellular environment. The anhydrous phosphate of PEP is hydrolyzed to release energy, and ATP can store around half of the available energy, which is equivalent to 30.5 kJ/mol (Anton et al., 1983).

7.1.1.3. Location of Glycolysis

In cells, the cytoplasmic matrix is where glycolysis takes place. This is likely because glycolysis is the oldest metabolic process before the emergence of intracellular organelles. A eukaryotic organism's glycolytically produced metabolites are transported into the mitochondria, fueling the aerobic tricarboxylic acid cycle (TCA cycle, also called citric acid or Krebs cycle) as well as electron transfer system processes (Brown et al., 1978).

7.1.1.4. Role of Glycolysis

The body's most basic technique for metabolizing sugar is called glycolysis. It helps produce both NADH that is utilized to make ATP in the electron transfer system, and ATP, the energy unit. The shunts for maintaining glycolysis are the fermentations of lactate & ethanol, which use the NADH produced to replicate NAD+, which is crucial for glycolysis. Only until the pyruvate from glycolysis has bonded with coenzyme A (CoA) to produce acetyl-CoA is it employed in the TCA cycle. Glycolysis has two functions. The production of ATP is the first function. Glycolysis produces ATP. However, the TCA cycle & the oxidative phosphorylation of acetyl-CoA also produce significant ATP (Connett et al., 1985).

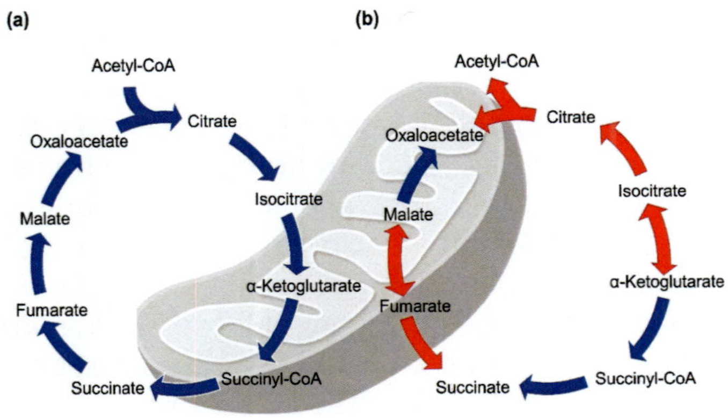

Figure 7.5. Tricarboxylic acid (Krebs or citric acid) cycle (Source: Kang et al., Creative Commons License).

Anaerobic conditions permit the production of just 2 moles of ATP per mole of glucose, while aerobic conditions permit the production of 38 moles of ATP per mole of glucose. The second function of this process is synthesizing intermediate metabolites, which serve as precursors for numerous metabolic pathways. Acetyl-CoA, for instance, is a precursor for the production of fatty acids. By fusing the acetyl residue and CoA, acetyl-CoA can be produced from proteins, carbohydrates, lipids, and pyruvate. Acetyl-CoA synthesized in this manner enters the TCA cycle, which has nine phases, located in the mitochondria matrix (Marcucci & Rumio, 2023).

The enzyme complex of the inner mitochondrial membrane carries out several reaction steps. Pyruvate, the final product of glycolysis, is converted into acetyl-CoA via decarboxylation & CoA. Acetyl-CoA can also be produced by amino acid metabolism & fatty acid b-oxidation. Lactate dehydrogenase converts pyruvate to lactate under anaerobic circumstances. Lactate is converted to pyruvate in people with acidosis, such as those with hypertension or diabetes. As a result, the lactate dehydrogenase reaction is reversible and is affected by oxygen levels in the body (Bajwa et al., 2016).

KEYWORD

Thyroid hormones are any hormones produced and released by the thyroid gland, namely triiodothyronine (T3) and thyroxine (T4). They are tyrosine-based hormones that are primarily responsible for regulation of metabolism.

7.1.2. Metabolism of Tricarboxylic Acid Cycle

During the TCA cycle, the acetyl group of acetyl-CoA is oxidized, which leads to the generation of two molecules of carbon dioxide as well as the reduction of 2 coenzymes, specifically NAD+ & flavine adenine dinucleotide (FAD+), to their reduced forms, (FADH2 & NADH+). It has a role in a wide variety of other processes, including the metabolism of amino acids, the urea cycle, glyconeogenesis, & breakdown of sugar. ATP is not directly synthesized throughout the TCA cycle, as it is during anaerobic glycolysis (also known as the Embden-Meyerhof-Parnas or EMP shunt). Aside from cyclically converting GTP into ATP, reduced coenzymes (FADH2 & NADH) produced will be utilized for ATP production via oxidative phosphorylation after the electron transfer mechanism. Sugar metabolism occurs in the cytoplasm & mitochondria of hepatic parenchymal cells. Skeletal muscle can continue to perform well in the absence of aerobic oxidation because glucose is a fuel for anaerobic glycolysis. This is because glucose is an energy source for the process. This makes it possible for an organism to survive in an environment with little oxygen (Araujo et al., 2012).

Glucose Regulation in Bloodstream: Food, glycogen degradation, and glyconeogenesis are the three blood glucose sources. During fasting, glycogen is destroyed in the liver & kidney before being released into the bloodstream. In contrast, muscle glycogen is used in the liver to provide muscle energy. However, glycogen amount which can be stored in the liver is limited, & glucose that can be produced from glycogen will quickly run out (Mailloux et al., 2007).

When fasting is maintained for a longer duration, glucose is produced via glyconeogenesis from glycerol. This process is triggered by the metabolism of several substances, including glycogenic amino acids (via the protein breakdown cycle) & lipids. Glucose is continuously synthesized and supplied even if the glyconeogenesis reaction is slightly slower than an equivalent decline in the blood sugar level (Krebs, 1954).

The flow of an entire metabolic pathway must be regulated to maintain the consistent chemical molecule supply required for pathways. It can be accomplished by regulating only one catalytic or 2 reactions in metabolism with "regulatory enzymes." Hormones regulate several regulatory enzymes activity. Monosaccharides, protein, lipid metabolism, and hormonal activity have all been documented. Insulin, growth hormone, glucagon, glucocorticoids, & adrenaline usually regulate glucose metabolism. (adrenaline) (Krebs, 1970).

KEYWORD

Electron transport system is a collection of proteins bound to the inner mitochondrial membrane and organic molecules, which electrons pass through in a series of redox reactions, and release energy.

Under stressful conditions, epinephrine (stress hormone) & cortisol levels rise, and anti-insulin hormones play a role in stress hyperglycemia pathogenesis. Hormones primarily regulate different intermediate metabolism between the fasting period & absorption phase. Insulin operates during absorption, while glucagon, epinephrine, thyroid hormone, growth hormone, adrenocortical hormone, and other hormones work while fasting. Insulin levels in the blood are reduced in type 1 diabetes patients due to congenital abnormalities in enzymes or cells involved in insulin secretion. Whereas, a decrease in insulin sensitivity is an aberrant response (receptor malfunction) & described as type 2 diabetes; the level of blood sugar in such patients tend to rise by using anti-diabetic medications such as acarbose (Guzman & Martin, 2009).

In the final step of aerobic respiration, an electron transfer system reaction occurs. It is also known as the respiratory chain, hydrogen transfer system, etc. Its goal is to create a proton concentration gradient between the interior and outside of a biomembrane. After that, enzymes that are responsible for the production of ATP use this proton concentration slope in order to produce ATP. (Rocha et al., 2010). Despite the fact that the majority of proteins that are part of the respiratory chain or coenzyme groups are embedded in biomembranes, cytochrome c is found on the surface of the membrane. Complex I is produced by the proton transported as NADH. MELAS is a condition caused by mutations in the mitochondrial electron transport pathway that affects 3000 newborns yearly. Electrons transported by $FADH_2$ are also transferred from complex II to ubiquinone (coenzyme Q) (Smith & Robinson, 2011).

By interacting with complex III, hydrogen ions ($2H^+$) in the reduced form of ubiquinone are transferred to intermembranous space while an electron is passed to complex III. The electron from complex III is transferred to complex IV via cytochrome *c*. Complex IV is responsible for the oxidation of the reduced form of cytochrome c & transfer of an electron to a molecule of oxygen (O2). One water molecule is formed when O2 reacts with two H+ in a matrix. Complexes I, III, & IV transport 10 H^+ to the intermembranous space via the space between complexes. The proton transfer gradient of H^+ concentration raises the membrane potential. Because O_2 is necessary for the final reaction of an electron transfer mechanism, aerobic respiration demands it (Amador et al., 2010).

KEYWORD

Alanine is an amino acid that is used to make proteins. It is used to break down tryptophan and vitamin B-6. It is a source of energy for muscles and the central nervous system.

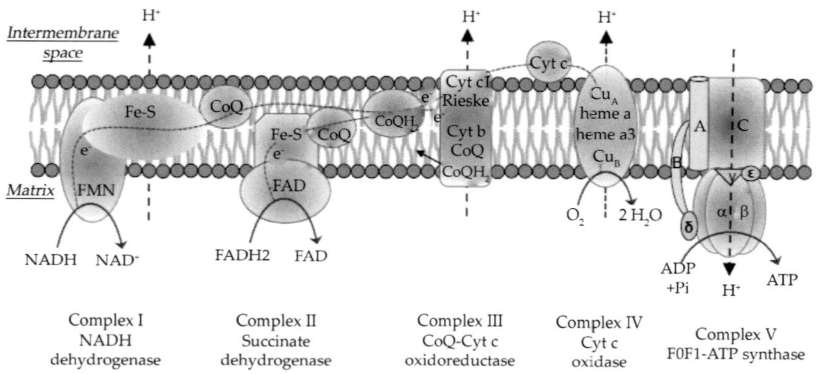

Figure 7.6. Adenosine triphosphate (ATP) synthesis & Electron transfer system &. I refers to the complex I (NADH dehydrogenase); II refers to the complex II (succinate dehydrogenase); III refers to the complex III; IV refers to the complex IV (cytochrome c oxidase; b refers to cytochrome b; c1 refers to cytochrome c1; c refers to cytochrome c; an refers to cytochrome a; FeS: iron-sulfur center; PT stands for phosphate ion transporter; Cu stands for cuprous protein; Q stands for ubiquinone (coenzyme Q); F0 stands for ATP synthase (F0 subunit); F1 stands for the F1 subunit (Source: Luo et al., Creative Commons License).

In the electron transport system, the potentially fatal toxin potassium cyanide acts as an inhibitor for complex IV. Complex IV deficiency causes Parkinson's disease and accelerates aging. Decreased complexity causes Alzheimer's disease I in the electron transport system. Complexes I & III deficiency may result in acidosis (Scagliola et al., 2020).

7.1.3. Glycogen Metabolism

7.1.3.1. Glycogen Synthesis

Glycogen is the body's reserve polysaccharide, primarily composed of hepatic glycogen. Glycogen is made in the liver as well as muscle. In the glycogen molecule, there's reduced sugar which exhibits reduction properties & many non-reducing residues which may not cause signal reduction (Preiss, 1984). There is a role for a residue in glycogen that does not reduce either when glycogen is cleaved or when glucose binds to glycogen. Glycogen synthase is a key enzyme in glycogen production. This enzyme facilitates the joining of one glucose molecule with glycogen. Glucose binding to glycogen forms a bond with a non-reducing residue (Stalmans et al., 1987).

KEYWORD

Distal convoluted tubule (DCT) is a short nephron segment, interposed between the macula densa and collecting duct. Even though it is short, it plays a key role in regulating extracellular fluid volume and electrolyte homeostasis.

7.1.3.2. Glycogen Degradation

Glycogen phosphorylase affects the glycogen breakdown reaction as it begins. As a result, glycogen can be used to produce glucose. Glycogen phosphorylase forms glycogen-1-phosphate by causing phosphorolytic cleavage into glycogen. However, when cleaving glucose from glycogen, a nonreducing terminal is eliminated. In addition, glycolysis can start right away if the glucose-1-phosphate that is synthesized from glycogen is converted into G6P by an enzyme called phosphoglucomutase (Fischer et al., 1971).

It should be noted that the consumption of ATP necessary for converting glucose to G6P is excluded when glycolysis proceeds directly. Therefore, 3 ATPs are produced if glycolysis uses glucose produced from glycogen. Glycogen synthase & glycogen phosphorylase activities regulate the metabolism of glycogen. Phosphorylation, which turns on glycogen phosphorylase and turns off glycogen synthase, is a key regulator of the metabolism (Kang et al., 2009).

7.1.3.3. Glucose-6-Phosphoric Acid Uses

The liver & muscles are the major locations of glycogen. Breakdown of liver glycogen results in the production of glucose, which affects the maintenance of blood sugar level, which is normally between 80 and 100 mg/dl. When glycogen in the muscle is broken down, the process of glycolysis starts, and a new source of energy is generated in the muscle cells (Kumler & Eiler, 1943).

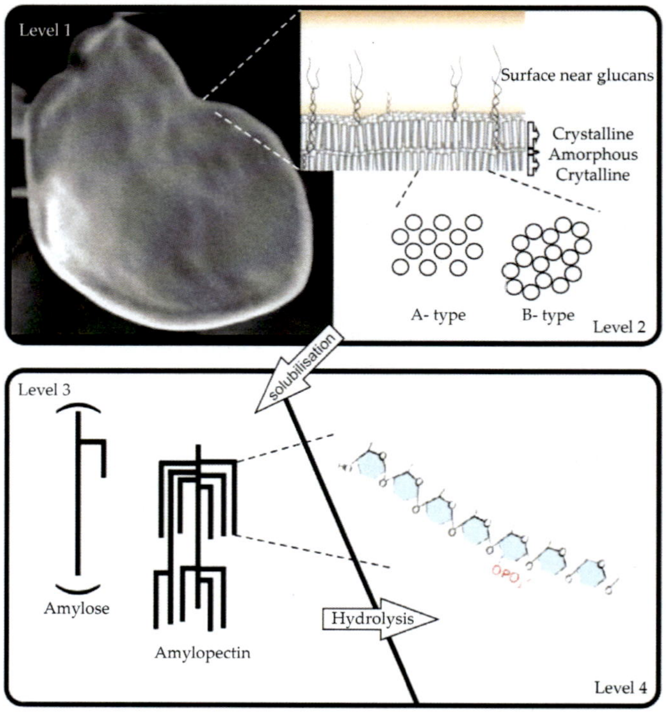

Figure 7.7. Glycogen hydrolysis (Source: Brust et al., Creative Commons License).

Although glucose-6-phosphatase contributes to the conversion of glucose-6-phosphoric acid to glucose, its large production has a strong activity in the liver and kidney. In contrast, the activity of glucose-6-phosphatase is minimal in muscles (Dev & Mishra, 1977). The G6P produced in muscles is therefore only used for energy production, as opposed to G6P produced in the liver, which is utilized for both glycolysis & glucose production, necessary to maintain blood sugar levels (Dietz, 1954).

7.1.4. Glyconeogenesis

The synthesis of sugars like glycogen & glucose from materials other than sugars is known as gluconeogenesis. Lactic acid to glucose conversion is one illustration. A glycolytic shunt increases the lactic acid to produce glucose through glyconeogenesis: lactic acid is converted into alanine, & delivered into the liver. Alanine is converted back into lactic acid in the liver & then glucose is produced through the process of gluconeogenesis (Markowitz, 1925). However, the kidney is where some glyconeogenesis takes place. Except for three one-way (irreversible) actions, glyconeogenesis can create glucose if enzymes are involved in the glycolytic shunt. The 3 reactions include: (1) Hexokinase converts glucose into G6P; (2) PFK-1 transforms F6P into F1,6BP; & (3) Pyruvate kinase transforms

PEP into pyruvate (Dyson, 1975). The use of additional enzymes, 1,6-bisphosphatase, fructose-& glucose-6-phosphatase, allows for the determination of reverse processes of (1) and (2). The final result, though, is more difficult. Pyruvate is first transported into mitochondria, catalyzed by pyruvate carboxylase, and transformed into oxaloacetate with CO2 and ATP. Oxaloacetic acid is first converted to malate, which is subsequently transported outside of mitochondria by the malate shuttle & converted back to oxaloacetate (Ryan & Radziuk, 1994). Lastly, the enzyme phosphoenolpyruvate carboxylase converts oxaloacetate to PEP. In this process, glucose is synthesized from two lactic acid molecules. For one molecule of glucose to be produced during glyconeogenesis, six ATP molecules are required (Jin et al., 2015).

KEYWORD

Acylcarnitines are esters of l-carnitine and fatty acids. They are a large class of metabolites that are members of the non-protein amino acid family.

7.1.5. Pentose Phosphate Pathway

The Pentose phosphate shunt is responsible for the production of nicotinamide adenine dinucleotide phosphate (NADPH) as well as D-ribose-5-phosphoric acid in reduced form. (Patra & Hay, 2014). NADPH is a coenzyme that functions in cytoplasm & connected to fatty acid production. Nucleic acid production is linked to D-ribose-5-phosphate. (DNA and RNA) (Vaseghi et al., 1999).

7.1.6. Blood Groups

The A blood type has the gene (A-transferase) that expresses A antigen, the AB blood group has genes (A-transferase & B-transferase) that expresses both A & B antigens and the B blood group has the gene (B-transferase) that expresses B antigen (Dean & Dean, 2005). H-transferase chemically arranges antigen H into antigens A and B. On chromosome 9, there is a gene that defines an individual's ABO blood type. On chromosome 19, the H-transferase precursor is converted to antigen H by the antigen H gene. The Bombay blood type does not result in the expression of this gene (Issitt & Anstee, 1975). Other blood group categorization systems exist besides the ABO system, including those based on the Rhesus factor, Lewis blood groups, & MN blood groups. For instance, around forty percent of persons in Japan have the blood type A, thirty percent have the blood group O, twenty percent have the blood group B, and ten percent have the AB blood group, whereas ninety percent or more of native South Americans have the blood group O (Hakomori, 1999). This number may even surpass ninety-nine percent depending on locality. In 1980, it was discovered that there was a link between the ABO blood group & sickness.

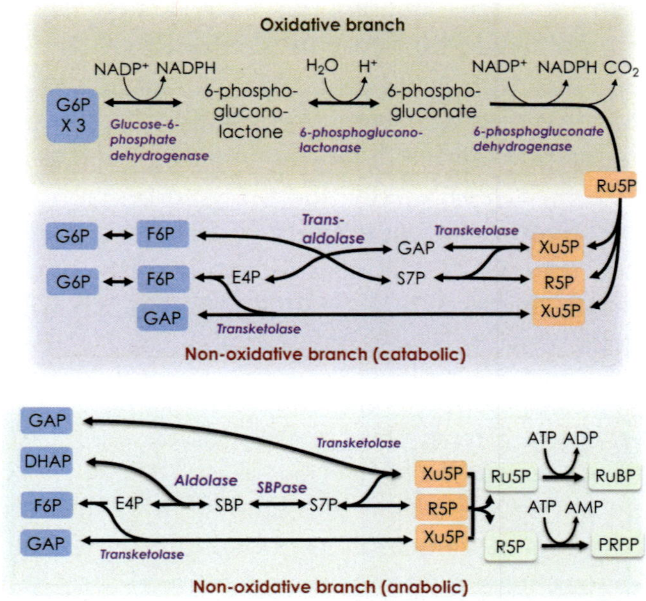

Figure 7.8. Schematic of the pentose phosphate shunt (Source: Sharkey et al., Creative Commons License).

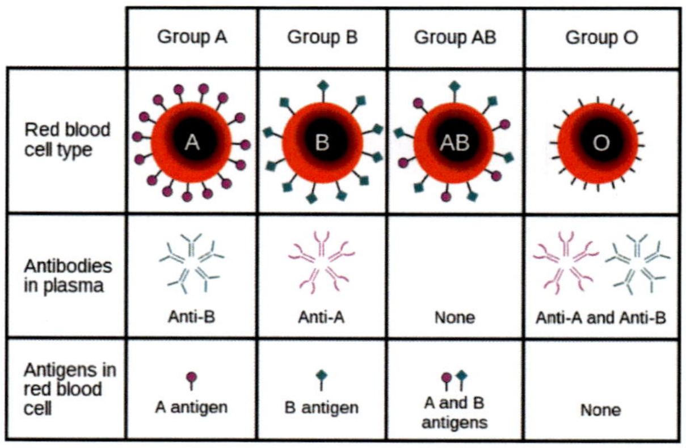

Figure 7.9. ABH and Lewis blood groups (Source: wikimedia, Creative Commons License).

There is a marginal relationship between Helicobacter pylori infection & ABO blood group. The Lewis secretory type, in contrast, has a close connection to *H. pylori* infection. The ABO blood group of a recipient may change after a bone marrow transplant. For instance, after receiving a bone marrow transplant, the deceased Ichikawa Danjuro XII's myelocytic leukemia patient's ABO blood group was altered (Springer & Horton, 1969).

7.1.7. Glycoprotein Synthesis

On the endoplasmic reticulum, post-translational modifications results in the formation of an asparagine-linked, mannose-rich sugar chain. It can subsequently be organized into a complex or O-linked sugar chain in the Golgi region (Sears & Wong, 1998). Glycosylphosphatidylinositol (GPI), a form of glycolipid present in the second type of glycoprotein, is also organized post-translationally on the endoplasmic reticulum. The GPI anchor connects the cell membrane in which the latter GPI glycoproteins are found. Alkaline phosphatase, cholinesterase, prion glycoprotein, & Tamm-Horsfall protein are only a few of the at least 150 varieties of GPI proteins found in the distal renal tubules. The process by which GPI is synthesized, the specifics of how GPI and proteins are linked, & physiological function of GPI anchor are all still unknown (Gleeson & Schachter, 1983).

GPI-dependent amidotransferase can produce a post-translational modification in the GPI anchor (Kuo & Lampen, 1974).

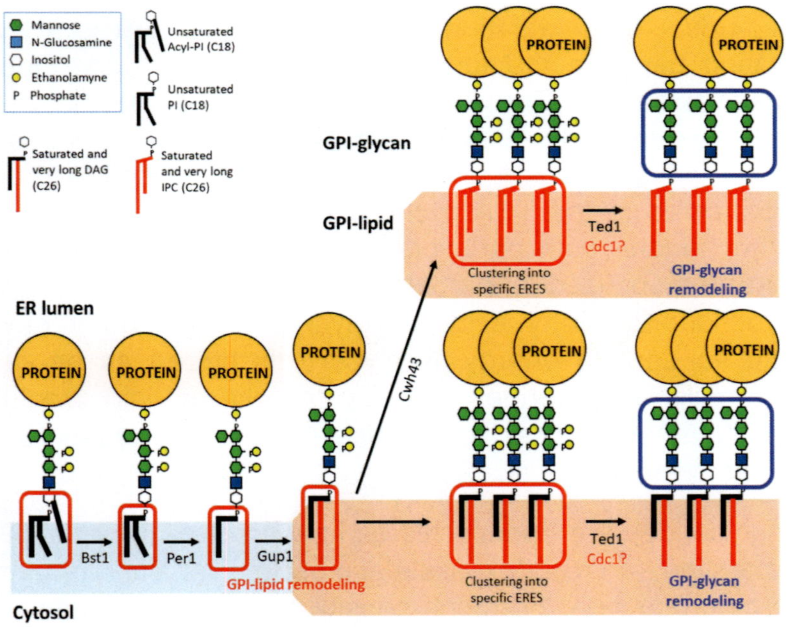

Figure 7.10. Illustration of how glycosylphosphatidylinositol (GPI)-glycoprotein is made in yeast (Source: Lopez et al., Creative Commons License).

It is possible that an amidotransferase is responsible for the removal of a hydrophobic amino acid sequence in the GPI glycoprotein precursor and its subsequent replacement by the GPI anchor; however, this enzyme has not yet been discovered. The paroxysmal hemoglobinuria (PNH or red diaper syndrome) aberrant degradation steps in the GPI constructional system's early phases significantly impact the complement fixation reaction at night. Red cell membranes are damaged by complement-mediated hemolysis in PNH, an acquired illness. Intravascular hemolysis, or the breakdown of red blood cells, occurs inside the blood artery

(Yang et al., 2007). Here, the complement response is mediated by a protein known as complement C3. CD34, CD55, and CD59 are examples of membrane proteins whose quality is controlled by the disease. In this disorder, there is either an absence of GPI anchors or a lack of synthesis of protein-linked GPI of the cell membrane. As a result, the red cell membrane is damaged, and regulation of C3 activation is lost. PNH is also caused by abnormalities in the phosphatidylinositol glycan class A gene (*PIG-A*), which is involved in GPI synthesis (Bennett & Wong, 2007).

KEYWORD

Flavin adenine dinucleotide is a redox-active coenzyme associated with various proteins, which is involved with several enzymatic reactions in metabolism.

7.2. LIPID METABOLISM

A metabolic process called b-oxidation oxidizes fatty acids and produces acetyl-CoA and fatty acyl-CoA, thioesters of fatty acids & CoA. Acetyl-CoA molecule is produced during the four repeated processes that make up b βoxidation, and acetyl-CoA is also the byproduct of β-oxidation of the fatty acid. It was known as β-oxidation because the fatty b position of acyl-CoA undergoes stepwise oxidation. The TCA cycle, followed by the electron transport system and β-oxidation, is the first of three phases in the metabolism of fatty acids (Lands, 1965). Acetyl-CoA is produced and connected to the TCA cycle, oxidizing to CO2. Peroxisomes and mitochondria both participate in β-oxidation. Long-chain fatty acids are metabolized by peroxisomes, and short-chain fatty acids by mitochondria. Fatty acids or glycerol in the body have the potential to deplete energy, & glycerol is stored as an ester in adipose cells. Triglycerides build up in cells as lipid droplets, so storage lipids don't dissolve or raise the osmotic pressure of the cytoplasm (Nguyen et al., 2008).

Triglycerides are transported as lipid droplets to organs (heart, small intestine, renal cortex, skeletal muscle, liver, etc.) for utilization in energy production. Hormone-sensitive lipase first makes its way to a lipid droplet's surface. Triglycerides can be hydrolyzed by lipase (Drackley, 2000). The muscle tissue transports hydrolyzed fatty acid via the bloodstream. The fatty acid transporter allows fatty acid separated from serum albumin to enter cells. Acetyl-CoA is produced due to the activation of fatty acids and b-oxidation after being taken in by cells. On the cytoplasmic side of the mitochondrial outer membrane is an enzyme called acyl-CoA synthase, which is also known as fatty acid thiokinase. This enzyme is responsible for activating the fatty acid that has been taken into the cell so that it can achieve stable C-C binding in the subsequent phase (Griminger, 1986):

Fatty acid + CoA + ATP → Fatty acyl-CoA + AMP + PP_i

There are two steps in this reaction. Fatty acyl adenylate and pyrophosphate are first synthesized when the carboxylate ion of fatty acid reacts with the phosphate of ATP. The nucleophilic attack of the acyl residue's carbon is then done by the thiol residue of CoA, producing fatty acyl-CoA & adenosine monophosphate (AMP).

A substance with high energy is fatty acyl-CoA. Carnitine is crucial to the fatty acyl shuttle because acyl-CoA can penetrate the mitochondrial inner membrane. Carnitine and fatty acyl-CoA bond briefly before producing fatty acid acylcarnitine (Williams, 2004).

Carnitine acyltransferase I, in the outer membrane of mitochondria, is responsible for catalyzing this process. It is currently unknown where the production of carnitine occurs, even though fatty acid acylcarnitine is supposed to be produced in intermembrane space & on the cytosolic side of the mitochondrial outer membrane. The acylcarnitine/carnitine transporter facilitates fatty acid acyl carnitine's diffusion across the membrane as it moves through the matrix. Carnitine acyltransferase II is an enzyme found in the mitochondrial inner membrane. It moves the fatty acyl group from carnitine to CoA to make acyl-CoA. The acylcarnitine/carnitine transporter transports free carnitine back to intramembrane space. Carnitine shuttle is the name of this fatty acyl-CoA transfer mechanism. Enzymes in the matrix oxidize fatty acyl-CoA that is transported into mitochondria (Maan et al., 2018).

The β-oxidation reaction, which removes two carbons from the carboxy terminus of the fatty acyl chain as an acetyl-CoA, comprises four repeating reaction stages, as was already explained. For instance, the even-numbered carbon atom palmitic acid (C16) is transformed into palmitoyl-CoA, which undergoes seven oxidations in the mitochondria to provide seven molecules of acetyl-CoA.. Additionally, 8 molecules of acetyl-CoA are formed in the entire reaction since two carbons are left over as acetyl-CoA at the unreduced end. Similar reactions occur when odd-carbon fatty acids are combined to form propionyl-CoA (CH_3CH_2CO-SCoA), which has three carbons and acetyl-CoA (Toth & Tchernof, 2000).

In addition to acyl-CoA mentioned above synthase of β-oxidation, the b-oxidation enzyme system also needs four more enzymes. These include acyl-CoA dehydrogenase (EC 1.3.99.3), enoyl-CoA hydratase (EC 4.2.1.17), acyl-CoA synthase (inner chain EC 6.2.1.2, long-chain EC 6.2.1.3), and b-hydroxy acyl-CoA dehydrogenase. (EC 1.1.1.35).

KEYWORD

Angiotensin-converting enzyme (ACE) inhibitors are medications that help relax the veins and arteries to lower blood pressure.

KEYWORD

Electrophoresis is a laboratory technique used to separate DNA, RNA or protein molecules based on their size and electrical charge. An electric current is used to move the molecules through a gel or other matrix.

Aryl-CoA dehydrogenase catalyzes the oxidation reaction, which is the first stage. A double bond is synthesized between the carbons of atoms a and b, producing *trans*-Δ^2-enoyl-CoA, in this step. The double bond is indicated through this symbol "." Mammals have 4 acyl-CoA dehydrogenase isozymes, each of which has a distinct action on acyl chains of different lengths. As a result of the formation of the double bond, an electron is transferred to the coenzyme of these enzymes known as flavine adenine dinucleotide (FAD), where it is then immediately transferred to the electron transport chain of mitochondria. A second FAD linked to hydrophilic electron transport flavoprotein (ETF) captures this electron (Harwood & Moore, 1989).

A double bond synthesized in the first stage is combined with H_2O in the second step, which is mediated by the enzyme enoyl-CoA hydratase, to create b-hydroxyacylCoA. (β-hydroxyacyl-CoA & 3-hydroxyacyl-CoA). Only the L-type is produced as this reaction proceeds in a stereoselective manner.

β-hydroxy acyl-CoA dehydrogenase catalyzes 3rd step, which results in β-ketoacyl-CoA. The sole substrate of this enzyme is L-type b-hydroxyacylCoA. NAD^+ is required for this reaction. After the electron has been acquired by NAD^+ in order to produce NADH, it is then transmitted to complex I of the electron transfer system. The process that weakens the stable C-C bond occurs in three stages (Xu et al., 2012).

The reaction caused by b-ketoacyl-CoA thiolase, also known as thiolase, b-keto thiolase, & acetyl-CoA acetyltransferase, occurs in the fourth stage. The thiolytic cleavage of b-ketoacyl-CoA & CoA by enzyme produces 2 carbon-shortened fatty acids, acyl-CoA, & acetyl-CoA. CoA's thiol residue (-SH) attacks the carbonyl carbon of β-ketoacyl-CoA in a nucleophilic manner, causing the a-b carbons to cleave (Sozio & Crabb, 2008).

The initial step in the subsequent b-oxidation uses fatty acyl-CoA with a shorter carbon chain as substrate. The reaction is continued until the entire fatty acyl chain has converted to acetyl-CoA. (or propinyl-CoA). To explain an equation palmitate can be used as an example. A significant amount of acetyl-CoA, a versatile molecule needed in many metabolic processes, is produced as the end product of β-oxidation. The following reactions involve acetyl-CoA (Frayn, 1993).

Arachidonic acid, an unsaturated fatty acid, makes eicosanoids such as leukotrienes (LTB_4) and prostaglandins (PGE_2, $PGF_{2a,}$ and

PGD_2). They play a role in generating reactive oxygen species (ROS), uterine contractions, childbirth, and sleep. The circadian rhythm, likewise regulated by clock genes like *Bmal1*, is used by edaravone to promote sleep by scavenging ROS. On an international journey, melatonin may be taken to reduce jet lag (Stumpf, 1960).

7.2.1. Steroid Hormones

The manufacture of sex hormones as well as other steroid hormones, including testosterone, estrogen, cortisone, & aldosterone, from cholesterol, is catalyzed by cytochrome cholesterol side-chain cleavage enzyme (P450scc). Reduced estradiol cause osteoporosis, decreased P450scc activity may cause hermaphroditism, and decreased testosterone may cause cryptorchidism & infertility in males (Shore & Shemesh, 2003). Cortisone has an anti-inflammatory impact and affects glyconeogenesis and skeletal metabolism. Angiotensin-converting enzyme (ACE), made in the kidney, converts angiotensin I into the molecule angiotensin II, which stimulates the release of the mineralocorticoid aldosterone, which raises blood pressure. An increase in aldosterone can bring on diabetes and acidosis (Lösel & Wehling, 2003).

The hormone corticosteroid produced by the adrenal cortex is involved in several physiological processes, including the regulation of inflammation, blood electrolytes, protein catabolism, glucose metabolism, & immunoreactivity. The hormones, also employed in medications, are released in response to numerous conditions, including stress and invasion. Mineralocorticoids and glucocorticoids are the hormones produced by the adrenal cortex (Gorski & Gannon, 1976). By regulating carbohydrates, fats, and proteins metabolism & inhibiting the formation of phosphatides, glucocorticoids act as anti-inflammatory drugs and reduce acidophilic material. The production of mineralocorticoids, which regulate the electrolyte and water balance, occurs mostly in the adrenal gland. Aldosterone and cortisol are the two common hormones. The cell membrane contains several steroid hormone receptors (R), & SR complex or its end product binds to a particular nucleic acid sequence in the nucleus. Then, specific proteins can be automatically synthesized (Beato, 1989).

As statin derivatives, ezetimibe and mevalotin selectively inhibit 3-hydroxy-3-methylglutaryl-CoA (HMG-CoA) reductase and the cholesterol transporter. Mammalian cell membranes need cholesterol for appropriate cell activity. Acetyl-CoA serves as the starting

KEYWORD

Microsomal triglyceride transfer protein (MTP) is a member of a group of proteins that are able to transfer of lipids between membranes. MTP is an 894 amino acid (97 kDa) protein that forms a heterodimer with protein disulfide isomerase (PDI); the two subunits are held together by non-covalent interactions.

material in the intracellular endoplasmic reticulum of cells during the creation of the lipid. The acyl ester of cholesterol in the blood is transported into low-density lipoprotein (LDL), which is then integrated into cells through receptor-mediated endocytosis, by inclusion into clathrin-coated pits, & then degraded by lysosomes (Wehling, 1997). Exogenous cholesterol is synthesized from bile acid & dietary sources & is absorbed through the intestinal duct. Endogenous cholesterol is mostly produced in the liver through mevalonic acid & squalene from acetyl-CoA. The amount of biosynthesis is modified to account for variations in the amount of exogenous cholesterol. For a person weighing 68 kg, the total quantity of cholesterol in their body is approximately 35 g. While most cholesterol is found in cell membranes, some are in the blood (Falkenstein et al., 2000).

TIPS

Lipid metabolism is like a complex dance, where enzymes and molecules move in harmony to orchestrate the storage and utilization of energy in the body

endogenous cholesterol production increases to about 800 mg per day, and the liver is responsible for producing 20–25% of the cholesterol that is circulating in the blood (Hartmann et al., 1998). This suggests the bloodstream contains 50% of the body's total cholesterol. The liver and skin have the highest cholesterol levels, respectively; no other organs have been shown to have it, except for the hippocampus region of the brain. Steroid hormones needed by organs other than the skin and liver are not biosynthesized in mammals, including humans (Siiteri et al., 1982).

It has been demonstrated that decreased testosterone production causes schizophrenia by reducing NADH metabolism (Liu et al., 2011).

De novo synthesis is used to create cholesterol, which is then carried from the liver to the plasma by a phospholipid complex that contains its esters. The form of cholesterol does not change while stored in the body. For instance, lipoprotein (LDL-cholesterol) in the body's midsection is a reservoir for cholesterol. When lysosomes hydrolyze cholesterol ester utilized in peripheral organs with phospholipids, cholesterol is released (Picard, 2006). Given that cholesterol is in such a dynamic condition and that the manufacture of cholesterol in the liver and through dietary absorption is essential, the cholesterolemia brought on by consuming too much cholesterol might be problematic. In addition to increased exogenous cholesterol from food, hypercholesterolemia is also brought on by decreased LDL-cholesterol receptors in peripheral organs. The consequence is a reduction in cholesterol uptake in familial hypercholesterolemia (FH) (Beato & Klug, 2000).

Additionally, aberrant lipid metabolism is believed to exist. Additionally, lifestyle diseases like metabolic syndrome, defined by increased blood pressure, high-density lipoprotein (HDL) cholesterol, HbA1c, and waist circumference, are gaining attention as complicated metabolic disorders of lipids & sugars (Giguère et al., 1988). Mice with non-alcoholic fatty liver disease were treated with MIYAIRI 588 (Miyarisan Pharmaceutical Co.) for elevated fatty acid levels (Seo et al., 2013).

7.2.2. Lipoprotein Metabolism

Blood plasma contains lipoproteins. Besides polar molecules like free fatty acids, lipids need to bond with protein as apoproteins to exist steadily in the plasma (Eisenberg & Levy, 1975). The many triglycerides and cholesterol required to maintain cells are found in the spherical lipoproteins that make up these particles (Ramasamy, 2014).

7.2.3. Lipoprotein Species

Table 4.2 lists the different types of lipoproteins and their main apolipoproteins. Using electrophoretic or supercentrifugal techniques, lipoproteins can be categorized. The classification of hyperlipidemia by the World Health Organization (WHO) provides the benchmark lipoprotein profile. The following categories can be used to categorize these lipoproteins: Chylomicrons, high-density lipoprotein (HDL), very high-density lipoprotein (VHDL), low-density lipoprotein (LDL), very low-density lipoprotein (VLDL), & intermediate-density lipoprotein (IDL) (Griffin, 1999).

When there are lipoprotein abnormalities, IDLs can be found. The ratio of cholesterol increases as a result of the way that chylomicron loses triglyceride when a VLDL particle's triglyceride is broken down and lost, & this LDL-cholesterol is known as remnant lipoprotein; it is equivalent to the IDL fraction. The LDL molecule known as small dense low-density lipoprotein (sd-LDL) is high in triglycerides (Dietschy et al., 1993). Using supercentrifugal or disc electrophoresis techniques, IDLs can be found in patients with advanced arteriosclerosis. Between VLDL and LDL, there is a band called IDL.. Cholesterol and triglycerides are contained in chylomicrons, lipoproteins with a density of less than 0.96 g/ml & ratio of roughly 1:10 (Kane et al., 2021). The proximal intestinal membrane's lipoprotein synthesizes lipids received from

KEYWORD

Cholesteryl ester transfer protein (CETP) is a hydrophobic plasma glycoprotein that mediates the transfer and exchange of cholesteryl ester (CE) and triglyceride (TG) between plasma lipoproteins, and also plays an important role in HDL metabolism.

the intestinal duct, which is then transported into the liver through the lymph vessel. The chylomicron, whose primary apoprotein is ApoB-48, is the lipoprotein that transports lipids. Chylomicronemia rises noticeably with LPL insufficiency. The main symptoms of abetalipoproteinemia, which are caused by decreased carrier particles of the lipophilic vitamins E & A, are peripheral nerve paralysis & night blindness. Abetalipoproteinemia is caused by the loss of microsome triglyceride transfer protein (MTP), mainly required for lipoprotein formation. IDL, also known as a remnant, is a lipoprotein with a 1.006–1.019 g/ml density produced when LPL hydrolyzes chylomicron and VLDL. Insulin resistance is a common medical history finding in patients diagnosed with metabolic syndrome. Patients with type III hyperlipemia with ApoE variation have lower LPL activity because the circulating IDL cannot connect to the receptors for VLDL, LDL, & lipoprotein receptor-related protein (LRP), which leads to a buildup in the bloodstream (Matsunaga et al., 2013).

KEYWORD

Tetrahydrofolic acid is produced from dihydrofolic acid by dihydrofolate reductase. This reaction is inhibited by methotrexate. It is converted into 5,10- methylenetetr-ahydrofolate by serine hydroxymethy-ltransferase.

IDL can be qualitatively recognized as a mid-band using the polyacrylamide gel electrophoresis technique.

In addition, the quantity of remnant-like particle cholesterol (RLP-C) can be determined with the help of an anti-ApoA-I antibody and an anti-ApoB-100 antibody, which are both included in the testing kit. It has been suggested to evaluate using the quantification of ApoB-48. LDL particles are 26–27 nm in size and have a density of 1.006–1.063 g/ml. LDL is referred to as "bad cholesterol" at times. The LDL receptor, which identifies ApoB & ApoE, is primarily responsible for transporting and metabolizing lipoprotein into the liver (Griffin, 1999). Familial hypercholesterolemia (FH) is the medical term for LDL receptor insufficiency; even in adolescence, total cholesterol content in FH homozygotes is at least 600 mg/dl, which results in severe arteriosclerosis-like ischemic heart disease. LDL loses its affinity for the LDL receptor by undergoing oxidation, denaturation, and glycation modifications. Then, fat-filled foam cells & thick atheromatous plaque are produced due to oxidized LDL being taken up by macrophage through scavenger receptor BI (LOX-1) in that situation (Curtiss & Edgington, 1976).

Since oxidative modification reduces the size of LDL and HDL, the blood vessel wall contributes to ischemic heart disease (Matsunaga et al., 2013). Small dense LDL (sd-LDL) and small dense HDL (sd-HDL) levels are higher in the blood of patients with hyperlipidemia. However fibrate-based medications can lower these triglyceride-rich particles. Healthy HDL has a density of 1.063-1.21

g/ml & an HDL particle diameter of 10-12.5 nm. ApoAI and ApoA-II make up the majority of the apoproteins in HDL particles (Tsutsumi et al., 2001). The "good" cholesterol HDL removes deposited cholesterol from vascular endothelium & is hypothesized to prevent arteriosclerosis by acting as a "reverse cholesterol transfer" system. Cholesteryl ester transfer protein (CETP) and reverse cholesterol transfer protein (RCTP) enable the transfer of neutral fat between VLDL & HDL & cholesterol between LDL & VLDL, respectively. By using RCTP, the liver can receive all the cholesterol. On the other hand, HDL that has been oxidized appears to lose its beneficial properties (Soine et al., 1982).

7.3. AMINO ACID METABOLISM

Transaminases are responsible for catalyzing the transfer of amino residues. Oxaloacetic acid is synthesized by the amino acids and aspartic acid from the TCA cycle . It similarly is broken down into oxaloacetic acid after being degraded by asparaginase & converting asparagine into aspartic acid. Proline, histidine, glutamine, and proline combine to form glutamic acid, which is then converted by glutamic acid dehydrogenases into 2-oxoglutaric acid (Miflin & Lea, 1977).

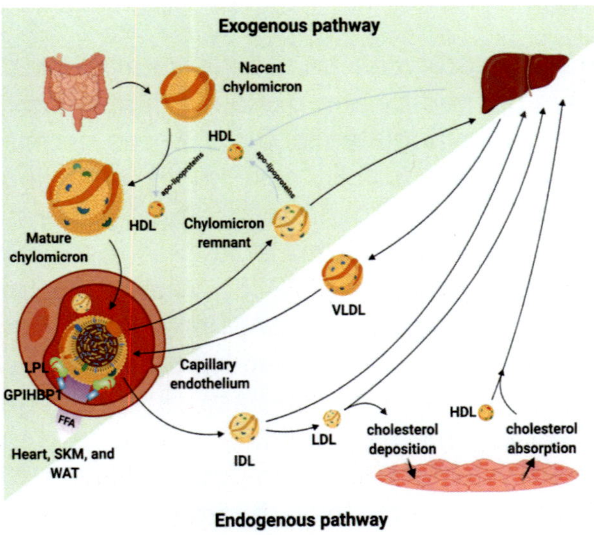

Figure 7.11. Illustration of metabolism of lipoproteins C: cholesterol; TG: triglyceride (neutral lipid); E: apoprotein LPL: lipoprotein lipase; E; LDL: low-density lipoprotein; VLDL: very low-density lipoprotein;HDL: high-density lipoprotein;IDL: intermediate-density lipoprotein; MTP: microsome triglyceride transfer protein; LCAT: lecithin cholesterol

KEYWORD

Branched-chain alpha-keto acid dehydrogenase (BCKDH) complex catalyzes the committed step of branched-chain amino acid catabolism, and its activity is regulated by the phosphorylation-dephosphorylation cycle. BCKDH kinase is responsible for inactivation of the complex by phosphorylation.

acyltransferase; HTGL: hepatic triglyceride lipase; ABCA1: ATP binding cassette A1; CETP: cholesteryl ester transfer protein (Source: Kumari et al., Creative Commons License).

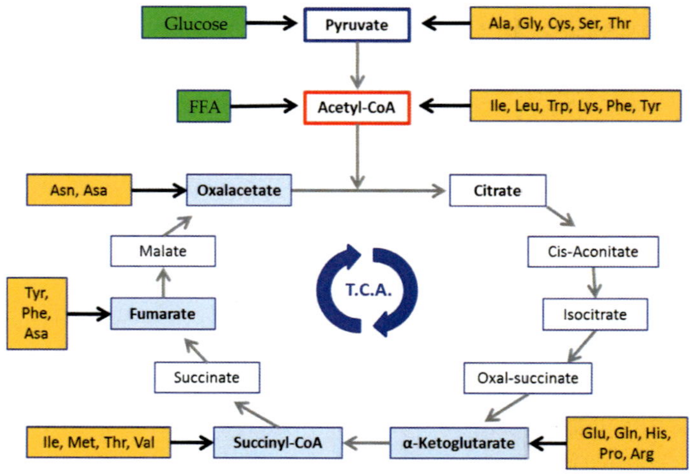

Figure 7.12. Illustration of Tricarboxylic acid cycle & amino acids: FFA = free fatty acids. Ala = alanine, Arg = arginine, Asa = aspartic acid, Asn = asparagine, Cys = cysteine, Gln = glutamine, Glu = glutamic acid, Gly = glycine, His = histidine, Ile = isoleucine, Leu = Leucine, Lys = lysine, Met = methionine, Phe = phenylalanine, Pro = proline, Ser = serine, Thr = threonine, Trp = tryptophan, Tyr = tyrosine, Val = valine. (Source: Pasini et al., Creative Commons License).

Glutaminase hydrolyzes glutamine into glutamic acid. The imidazole ring is opened by non-oxidative deamination of histidine, resulting in N-formiminoglutamic acid, & the formimino residue is transferred to tetrahydrofolic acid (THF), resulting in glutamic acid & 5-homoimino tetrahydrofolic acid. (5-hormimino THF). Through the creation of glutamic acid 5-semialdehyde, arginine, and proline are produced (Felig, 1975). When methionine interacts with ATP, S-adenosylmethionine (SAM or AdoMet) is synthesized. Epinephrine (adrenaline) & Phosphatidylethanolamine are changed into norepinephrine (noradrenaline) & phosphatidylcholine by SAM, a potent methyl residue donor. SAM will act as S-adenosylhomocysteine if the substrate is given with a methyl residue. S-adenosylhomocysteine hydrolyzes to form adenosine and homocysteine, which are then changed into cystathionine or methionine (by replicating the methyl residue). (by binding to serine) (Wu, 2009).

An imbalance between the production & breakdown of methylhomocysteine results in high homocysteinemia. High homocysteinemia first appears in the newborn, which results in cardiovascular illness, neural tube abnormalities, and anencephaly. Vitamin B12, B6 & folic acid consumption during pregnancy can prevent genetic homocysteinemia in newborns (Eagle, 1959). Leucine, isoleucine, & valine are three branched-chain amino acids (Morot et al., 2001).

7.3.1. Leucine

A deficiency in an enzyme known as branched-chain a-keto acid dehydrogenase (BCKDH) is the root cause of a disorder known as maple syrup urine disease (MSUD). Because of an enzyme deficiency, an excessive amount of branched a-keto acid is released in the urine, which gives it a scent similar to that of maple syrup (Kobe & Kajava, 2001). Because MSUD is an autosomal recessive disorder, if a patient isn't given treatment as soon as they are diagnosed, they will quickly pass away from the disease (Görbitz & Dalhus, 1996).

KEYWORD

Urea cycle is a cycle of biochemical reactions that produces urea $(NH_2)_2CO$ from ammonia. Animals that use this cycle, mainly amphibians and mammals, are called ureotelic.

7.3.2. Isoleucine and Valine

Branched amino acid transaminase, acyl-CoA dehydrogenase, branched a-keto acid dehydrogenase, & enoyl-CoA hydratase are four enzymes that catalyze the first four steps in the breakdown of isoleucine. These enzymes are also responsible for the metabolization of valine in the same manner. Acetyl-CoA is produced in addition to propionyl-CoA in the breakdown route of 3-hydroxy-2-methylbutyryl-CoA obtained from isoleucine (Yu et al., 2021)

7.3.3. Phenylalanine & Tyrosine

Tyrosine is used to metabolize phenylalanine to produce acetoacetic acid and fumaric acid. Both epinephrine and norepinephrine are produced from the tyrosine metabolite DOPA. These neurotransmitters play a causal role in developing schizophrenia and Parkinson's disease. The b-blocker chlorpromazine is useful in treating them, although osteoporosis is one of its side effects. When chlorpromazine is used to treat schizophrenic patients, the doctor should suggest taking active vitamin D (Matthews, 2007).

Because of the oxidation of homogentisic acid, patients with alcaptonuria have dark urine. Phenylketonuria is caused by an error in phenylalanine metabolism caused by a loss or dysfunction of phenylalanine hydroxylase, that converts phenylalanine to tyrosine.

The phenylalanine concentration in the blood is increased due to this issue. A patient with phenylalaninemia (excess phenylalanine) creates more phenyllactic acid, phenylpyruvic acid, & phenylacetic acid, culminating in phenylketonuria & schizophrenia (Moja et al., 1996). A shortage of the enzyme tyrosinase leads to albinism and a deficiency in the neurotransmitter dopamine, both of which can contribute to symptoms of schizophrenia (Tessari et al., 2010).

7.4. UREA CYCLE

KEYWORD

Cytoplasm is the gelatinous liquid that fills the inside of a cell. It is composed of water, salts, and various organic molecules. Some intracellular organelles, such the nucleus and mitochondria, are enclosed by membranes that separate them from the cytoplasm.

The urea cycle (known as the ornithine cycle) produces urea from ammonia. Hans Krebs & Kurt Henseleit discovered this metabolism in 1932 (Morris, 2002). This cycle is seen in practically every land vertebrate, with reactions occurring in the mitochondria of hepatic cells & cytosol (Leonard & Morris, 2002).

7.5. NUCLEIC ACID METABOLISM

7.5.1. Purine Metabolism

Using a new synthesis shunt de novo cycle that is based on 5-phosphoribosyl-1-pyrophosphate (PRPP), purine bases can be synthesized from glycine, glutamine, and other amino acids (Zimmermann, 1996). Inosine monophosphate (IMP), the precursor of guanine or adenine, is produced as a result of this process. PRPP amidotransferase adds one amino residue from glutamic acid (Moriwaki et al., 1999).

7.5.2. Nucleotide Decomposition

Deamination by adenosine deaminase converts adenosine to inosine, that's then transformed into hypoxanthine via dephosphorylation & subsequent deribosization. Uric acid is produced when hypoxanthine is oxidized. In general, the uric acid level is 7.0 mg/dl soluble in bodily fluid. Gout develops when blood concentration exceeds this level (Yang et al., 2015).

7.5.3. Gout

It is a type of metabolic syndrome. Because of a deficiency in purine metabolism, uric acid accumulates in the organs of this disease. The joints may accumulate uric acid. People should limit their consumption of meat, pork, and fowl. Draft beer raising uric acid concentrations in AMP and IMP (Emmerson, 1996).

Hypersynthetic gout or gout caused by a partial lack of hypoxanthine-guanine phosphoribosyltransferase (HGPRT) deficiency, commonly known as Lesch-Nyhan syndrome or Kelley-Seegmiller syndrome, are both types of hereditary gout. Both of these conditions are inherited. HGPRT participates in the purine salvage pathway, that helps the body restore purine. Purine body biosynthesis is enhanced in X-linked recessive hereditary gout. Only boys are affected by the syndrome, marked by mental disorders & self-injurious conduct. Despite being used to treat gout, the urate synthesis inhibitor urinome may also reduce uric acid reabsorption in renal tubules through peroxisome proliferator-activated receptor-a. (PPARa) (Kunishima et al., 2007).

Adenine is converted to AMP by the enzyme adenosine phosphoribosyltransferase (APRT). A rare syndrome characterized by autosomal recessive inheritance, APRT deficiency, is associated with purine salvage shunt urolithiasis & renal insufficiency caused by DHA.

In acute gouty arthritis, the metatarsophalangeal joint develops arthritis. After a gout episode, the patient's recovery is prolonged. The typical gout attack lasts for around two weeks and happens at night and in the morning when body temperatures drop. An increased erythrocyte sedimentation rate & greater C-reactive protein levels can be used to identify the inflammation.

Did you Know?

Protein biosynthesis is a highly complex and energy-intensive process that involves the precise coordination of hundreds of molecular machines and enzymes, with the average human cell synthesizing around 1 million proteins per second.

7.5.4. Partial Deficiency of PRPP Enzymes

In some circumstances, a loss of the enzyme that transforms adenosine into inosine results in immune insufficiency.

- PRPP + Adenine → (Adenine phosphoribosyltransferase)→ AMP + PP_i
- PRPP + Hypoxanthine → (Hypoxanthine – Guaninephosphoribosyltransferase)→ IMP + PP_i
- PRPP + Guanine → (Hypoxanthine – Guaninephosphoribosyltransferase)→ GMP + PP_i

Researchers tried to use retroviral vectors to deliver the normal adenosine deaminase (ADA) gene to patients as part of gene therapy (Kelley et al., 1967).

7.6. PROTEIN BIOSYNTHESIS

The method through which a cell produces protein is called protein biosynthesis. Despite the possibility that this phrase solely refers to translation in a very specific sense, in most cases, it refers to complex series of steps involved in the production of proteins, beginning with transcription & continuing with translation.

KEYWORD

N-Formylmethionine is a derivative of the amino acid methionine in which a formyl group has been added to the amino group.

Messenger ribonucleic acid (mRNA) for protein is synthesized through transcription from genomic DNA. The nucleotide sequence of the DNA double strands' sense strand, also known as the coding strand, is used for this. These three actions comprise the transcription process. RNA polymerase converts DNA into RNA by binding to a specific DNA sequence designating a transcription beginning site. A DNA chain, known as a promoter, will separate if RNA polymerase binds to this sequence. A ribonucleotide chain is synthesized as RNA polymerase proceeds along with a template or an antisense strand. A sense strand creates the complementary sequence, not another sense strand. The RNA is then synthesized using the DNA complement, or antisense strand. If polymerase runs out of energy, modification is carried out to speed up mRNA transit within cells. A poly(A) tail & cap structure are respectively attached to the 5' ends of the mRNA to prevent it from being degraded. Splicing in eukaryotes is carried out by an enzyme known as a spliceosome. This step involves cutting out the extra intron and joining the exon to create mature mRNA. After then, a pore in the nuclear membrane allows the mature mRNA to be synthesized to enter the cytoplasm (Merrick & Nyborg, 2000).

7.6.1. Initiation

The ribosome splits into two distinct subunits before protein synthesis begins. A complex of five different proteins is put together, including the small subunit (30S), mRNA, an initiating factor, formyl methionyl-transfer RNA, & guanosine triphosphate to start the protein synthesis process. (GTP). By building the sequence AGGAGGU close to the AUG codon, it is aligned to RNA. The particular sequence AGGAGGU (ribosome junction position), which is located close to the beginning codon AUG, & complementary sequence, which is located close to the 3' ends of 16S ribosomal RNA (rRNA) molecule, form a base pair that determines the binding location of the 30S subunit to mRNA. The big subunit (the 50S subunit) attaches to the 30S initiation complex after it has formed, creating the 70S initiation complex. The A (aminoacyl) & P (peptidyl) sites are 2 transfer RNA (tRNA) binding sites in the 50S subunit (Jolley et al., 2001).

7.6.2. Extension

When N-formylmethionine (fMet)-tRNA combines with site P, a tRNA molecule with an anticodon that interacts with the codon of a neighboring initiating codon can join with the A site of the 70S initiation complex. This occurs because the anticodon of the tRNA molecule is complementary to the codon of the neighboring initiating codon (Greene, 1974).

When an acylated tRNA strand is brought into close proximity with an elongation factor at an A site, the result is peptide bond formation between amino acid that has been combined with fMet & tRNA that is part of the peptidyltransferase enzyme complex. In that situation, site P of fMet-tRNA is cut off. The tRNA in the P site of the ribosome which split with an amino acid is released. Following, 2-peptidyl-tRNA transfers from the A site to the P site. These three nucleotides are followed by mRNA as it travels, & resulting codon appears in the A site. This motion is referred to as a translocation (Bull et al., 2004).

7.6.3. Termination

If the peptide chain is extended and the codon (UAG, UAA, or UGA) is reached from start to finish, no acylated tRNA can attach to the A site, & chain extension is stopped. The finished peptide chain exits the ribosome due to the termination factor, that detects these stop signals. The ribosome complex separates into its two components to be employed once more for producing fresh protein (Dalal & Mallows, 1988).

A particular enzyme removes the formyl group when the finished peptide chain leaves the ribosome, and in many circumstances, it also removes methionine at the end of the N-terminal residue. One mRNA molecule joins with several ribosome parts to effectively create proteins (Raviglione & Uplekar, 2006).

A polysome, which is the ribosome complex coupled to the mRNA, can manufacture a 100-residue polypeptide in just 5 seconds.

The other codon's anticodon-corresponding acylated tRNA molecule attaches to the A site, causing the peptide chain to lengthen. A freshly produced protein will undergo cotranslational modification and folding. Regarding the folding of a polypeptide chain, most proteins can be folded post-translationally in what is known as the secondary and tertiary structures of proteins. Functional groups are also added post-translationally, leading to some proteins' active conformation. Examples of these functional group additions include the creation of an acetyl residue, disulfide bond, phosphate residue, lipids, & a sugar moiety. In addition, signal peptidase is responsible for the cleavage of the hydrophobic polypeptide chain, also known as the signal peptide, located at the N-terminal amino acid sequence (Bertram et al., 2000).

ACTIVITY 7.1

John is a 45-year-old man who has been experiencing constant fatigue and unexplained weight gain for several months. His doctor suspects that he may have a metabolic disorder affecting his body's ability to properly break down and utilize nutrients.

- What are some possible metabolic pathways that could be disrupted in John's body, and how might they be affecting his symptoms?
- What tests could his doctor perform to identify the specific metabolic disorder affecting John, and what treatments might be recommended once a diagnosis is made?
- How might John's diet and lifestyle need to change in order to support proper functioning of his metabolic pathways and improve his symptoms?

SUMMARY

The chapter begins by explaining the basics of sugar metabolism, including glycolysis, the TCA cycle, and glycogen metabolism. The glycolytic pathway is a crucial energy-generating process, while the TCA cycle is responsible for the oxidation of acetyl-CoA and producing energy intermediates. Glycogen metabolism, on the other hand, involves the storage and breakdown of glucose in the liver and muscles.

The process of gluconeogenesis is explained in detail, which is the conversion of non-carbohydrate sources into glucose. This process is important in maintaining blood glucose levels during fasting or starvation. The role of the pentose phosphate pathway is discussed, which generates NADPH and ribose-5-phosphate, essential for the biosynthesis of nucleotides and fatty acids.

The chapter also covers the different blood groups and their importance in blood transfusions. Blood groups are determined by the presence or absence of specific antigens on red blood cells, which can cause an immune reaction during transfusion.

The basics of lipid metabolism are explained, including the synthesis and breakdown of lipids and the role of lipoproteins in transporting lipids in the blood. The synthesis and functions of steroid hormones are discussed. Steroid hormones are derived from cholesterol and play a critical role in regulating various physiological processes in the body, including metabolism, growth, and reproduction. Finally, the chapter explains the importance of amino acid metabolism, including the roles of leucine, isoleucine, valine, phenylalanine, and tyrosine. Amino acids are important building blocks of proteins, and their metabolism is essential for energy production and synthesis of non-protein molecules such as neurotransmitters and nucleotides.

REVIEW QUESTIONS

1. What is glycolysis and what are its products?
2. Describe the process of gluconeogenesis and explain its importance in the body.
3. What is the pentose phosphate pathway and what is its role in the body?
4. How blood groups determined and what are is their importance in blood transfusions?
5. What is lipid metabolism and what are lipoproteins?
6. Describe the synthesis and functions of steroid hormones.
7. What is amino acid metabolism and what are the roles of leucine, isoleucine, valine, phenylalanine, and tyrosine?

MULTIPLE CHOICE QUESTIONS

1. **Which of the following is a product of glycolysis?**
   a. Acetyl-CoA
   b. Pyruvate
   c. FADH2
   d. NADPH
2. **What is the process by which glucose is synthesized from non-carbohydrate sources?**
   a. Glycolysis
   b. Gluconeogenesis
   c. Pentose phosphate pathway
   d. Urea cycle
3. **Which of the following is a role of the pentose phosphate pathway?**
   a. ATP production
   b. NADPH and ribose-5-phosphate generation
   c. Fatty acid synthesis
   d. Glycogen breakdown
4. **Blood groups are determined by:**
   a. Antigens on the surface of red blood cells
   b. Antibodies in the plasma
   c. Both antigens and antibodies
   d. None of the above
5. **Which of the following is a complex of lipids and proteins that transports lipids in the blood?**
   a. Steroid hormone
   b. Lipoprotein
   c. Glycogen
   d. Pentose phosphate
6. **Which of the following is a hormone derived from cholesterol that acts as a signaling molecule in the body?**
   a. Glucagon
   b. Insulin
   c. Cortisol
   d. All of the above

7. Which of the following amino acids is essential?

a. Leucine

b. Glycine

c. Aspartic acid

d. None of the above

8. What is the process by which ammonia is converted into urea for excretion in urine?

a. Glycolysis

b. Gluconeogenesis

c. Urea cycle

d. Protein biosynthesis

9. What is the process by which purines are synthesized and broken down in the body?

a. Glycolysis

b. Gluconeogenesis

c. Purine metabolism

d. Protein biosynthesis

10. What is the process by which amino acids are linked together to form a protein?

a. Glycolysis

b. Gluconeogenesis

c. Urea cycle

d. Protein biosynthesis

Answers to Multiple Choice Questions

1. (b)	2. (b)	3. (b)	4. (a)	5. (b)
6. (c)	7. (a)	8. (c)	9. (c)	10. (d)

REFERENCE

1. Amador-Noguez, D., Feng, X. J., Fan, J., Roquet, N., Rabitz, H., & Rabinowitz, J. D. (2010). Systems-level metabolic flux profiling elucidates a complete, bifurcated tricarboxylic acid cycle in *Clostridium acetobutylicum*. *Journal of Bacteriology*, *192*(17), 4452-4461.

2. Anton, D. L., Hedstrom, L., Fish, S. M., & Abeles, R. H. (1983). Mechanism of enolpyruvylshikimate-3-phosphate synthase exchange of phosphoenolpyruvate with solvent protons. *Biochemistry*, *22*(25), 5903-5908.
3. Araujo, W. L., NUNES-NESI, A. D. R. I. A. N. O., Nikoloski, Z., Sweetlove, L. J., & Fernie, A. R. (2012). Metabolic control and regulation of the tricarboxylic acid cycle in photosynthetic and heterotrophic plant tissues. *Plant, Cell & Environment*, *35*(1), 1-21.
4. Arutyunova, E. I., Danshina, P. V., Domnina, L. V., Pleten, A. P., & Muronetz, V. I. (2003). Oxidation of glyceraldehyde-3-phosphate dehydrogenase enhances its binding to nucleic acids. *Biochemical and Biophysical Research Communications*, *307*(3), 547-552.
5. Ayna, A., & Moody, P. C. (2021). Crystal structures of a dual coenzyme specific glyceraldehyde-3-phosphate dehydrogenase from the enteric pathogen Campylobacter jejuni. *Journal of Molecular Structure*, *1242*, 130820.
6. Bajwa, G., DeBerardinis, R. J., Shao, B., Hall, B., Farrar, J. D., & Gill, M. A. (2016). Cutting edge: Critical role of glycolysis in human plasmacytoid dendritic cell antiviral responses. *The Journal of Immunology*, *196*(5), 2004-2009.
7. Beato, M. (1989). Gene regulation by steroid hormones. *Cell*, *56*(3), 335-344.
8. Beato, M., & Klug, J. (2000). Steroid hormone receptors: an update. *Human Reproduction Update*, *6*(3), 225-236.
9. Bennett, C. S., & Wong, C. H. (2007). Chemoenzymatic approaches to glycoprotein synthesis. *Chemical Society Reviews*, *36*(8), 1227-1238.
10. Bertram, G., Bell, H. A., Ritchie, D. W., Fullerton, G., & Stansfield, I. (2000). Terminating eukaryote translation: domain 1 of release factor eRF1 functions in stop codon recognition. *RNA*, *6*(9), 1236-1247.
11. Blake, C. C. F., & Rice, D. W. (1981). Phosphoglycerate kinase. *Philosophical Transactions of the Royal Society of London. B, Biological Sciences*, *293*(1063), 93-104.
12. Bonadonna, R. C., Bonora, E., Prato, S. D., Saccomani, M. P., Cobelli, C., Natali, A., ... & Gulli, G. (1996). Roles of glucose transport and glucose phosphorylation in muscle insulin resistance of NIDDM. *Diabetes*, *45*(7), 915-925.
13. Brown-Woodman, P. D., Mohri, H., Mohri, T., Suter, D. A. I., & White, I. G. (1978). Mode of action of α-chlorohydrin as a male anti-fertility agent. Inhibition of the metabolism of ram spermatozoa by α-chlorohydrin and location of block in glycolysis. *Biochemical Journal*, *170*(1), 23-37.
14. Bull, N. H., Cote, L. S., Warner, P. D., & McKinnie, M. (2004). Is extension relevant for the 21st century?. *The Journal of Extension*, *42*(6), 3.
15. Connett, R. J., Gayeski, T. E., & Honig, C. R. (1985). Energy sources in fully aerobic rest-work transitions: a new role for glycolysis. *American Journal of Physiology-Heart and Circulatory Physiology*, *248*(6), 922-929.

16. Curtiss, L. K., & Edgington, T. S. (1976). Regulatory serum lipoproteins: regulation of lymphocyte stimulation by a species of low density lipoprotein. *The Journal of Immunology*, *116*(5), 1452-1458.
17. Dalal, S. R., & Mallows, C. L. (1988). When should one stop testing software?. *Journal of the American Statistical Association*, *83*(403), 872-879.
18. Dean, L., & Dean, L. (2005). *Blood Groups and Red Cell Antigens* (Vol. 2).pp 56-140
19. Debarbouille, M., Gardan, R., Arnaud, M., & Rapoport, G. (1999). Role of BkdR, a transcriptional activator of the SigL-dependent isoleucine and valine degradation pathway in *Bacillus subtilis*. *Journal of Bacteriology*, *181*(7), 2059-2066.
20. Dev, B., & Mishra, G. C. (1977). Effect of decerebration on the enzymatic system of stomach and intestine of Indian cattle leech, Poecilobdella granulosa. *Zeitschrift für Parasitenkunde*, *52*, 97-101.
21. Dietschy, J. M., Turley, S. D., & Spady, D. K. (1993). Role of liver in the maintenance of cholesterol and low density lipoprotein homeostasis in different animal species, including humans. *Journal of Lipid Research*, *34*(10), 1637-1659.
22. Dietz, A. A. (1954). Biochemical Preparations, Vol. 3. Esmond E. Snell, Ed. New York: Wiley; London: Chapman & Hall, 1953. 128 pp. *Science*, *119*(3093), 482-482.
23. Dipietro, D. L., Sharma, C., & Weinhouse, S. (1962). Studies on glucose phosphorylation in rat liver. *Biochemistry*, *1*(3), 455-462.
24. Drackley, J. K. (2000). Lipid metabolism. *Farm animal metabolism and nutrition*, *1*, 97-119.
25. Dyson, R. D., Cardenas, J. M., & Barsotti, R. J. (1975). The reversibility of skeletal muscle pyruvate kinase and an assessment of its capacity to support glyconeogenesis. *Journal of Biological Chemistry*, *250*(9), 3316-3321.
26. Eagle, H. (1959). Amino acid metabolism in mammalian cell cultures. *Science*, *130*(3373), 432-437.
27. Eisenberg, S., & Levy, R. I. (1975). Lipoprotein metabolism. *Advances in Lipid Research*, *13*, 1-89.
28. Emmerson, B. T. (1996). The management of gout. *New England Journal of Medicine*, *334*(7), 445-451.
29. Enyedi, Z. (2005). The role of agency in cleavage formation. *European Journal of Political Research*, *44*(5), 697-720.
30. Falkenstein, E., Tillmann, H. C., Christ, M., Feuring, M., & Wehling, M. (2000). Multiple actions of steroid hormones—a focus on rapid, nongenomic effects. *Pharmacological Reviews*, *52*(4), 513-556.
31. Felig, P. (1975). Amino acid metabolism in man. *Annual Review of Biochemistry*, *44*(1), 933-955.
32. Fischer, E. H., Heilmeyer Jr, L. M., & Haschke, R. H. (1971). Phosphorylase and the control of glycogen degradation. In *Current Topics in Cellular Regulation* (Vol. 4, pp. 211-251).

33. Ford, R., & Jennings, W. (2020). The changing cleavage politics of Western Europe. *Annual Review of Political Science*, *23*(1), 295-314.
34. Frayn, K. N. (1993). Insulin resistance and lipid metabolism. *Current Opinion in Lipidology*, *4*(3), 197-204.
35. Giguère, V., Yang, N. A., Segui, P., & Evans, R. M. (1988). Identification of a new class of steroid hormone receptors. *Nature*, *331*(6151), 91-94
36. Gleeson, P. A., & Schachter, H. (1983). Control of glycoprotein synthesis. *Journal of Biological Chemistry*, *258*(10), 6162-6173.
37. Görbitz, C. H., & Dalhus, B. (1996). Redetermination of L-Leucine at 120K. *Acta Crystallographica Section C: Crystal Structure Communications*, *52*(7), 1754-1756.
38. Gordon, R. D., Iwatsuki, S., Esquivel, C. O., Tzakis, A., Todo, S., & Starzl, T. E. (1986). Liver transplantation across ABO blood groups. *Surgery*, *100*(2), 342-348.
39. Gorski, J., & Gannon, F. (1976). Current models of steroid hormone action: a critique. *Annual Review of Physiology*, *38*(1), 425-450.
40. Greene, C. (1974). An extension of Schensted's theorem. *Advances in Mathematics*, *14*(2), 254-265.
41. Griffin, B. A. (1999). Lipoprotein atherogenicity: an overview of current mechanisms. *Proceedings of the Nutrition Society*, *58*(1), 163-169.
42. Griminger, P. (1986). Lipid metabolism. *Avian Physiology*, 345-358.
43. Guzman, M. I., & Martin, S. T. (2009). Prebiotic metabolism: production by mineral photoelectrochemistry of α-ketocarboxylic acids in the reductive tricarboxylic acid cycle. *Astrobiology*, *9*(9), 833-842.
44. Hakomori, S. I. (1999). Antigen structure and genetic basis of histo-blood groups A, B and O: their changes associated with human cancer. *Biochimica et Biophysica Acta (BBA)-General Subjects*, *1473*(1), 247-266.
45. Hartmann, S., Lacorn, M., & Steinhart, H. (1998). Natural occurrence of steroid hormones in food. *Food Chemistry*, *62*(1), 7-20.
46. Harwood, J., & Moore Jr, T. S. (1989). Lipid metabolism in plants. *Critical reviews in plant sciences*, *8*(1), 1-43.
47. Hinckelmann, M. V., Virlogeux, A., Niehage, C., Poujol, C., Choquet, D., Hoflack, B., ... & Saudou, F. (2016). Self-propelling vesicles define glycolysis as the minimal energy machinery for neuronal transport. *Nature Communications*, *7*(1), 13233.
48. Hodgson, D. M., & Petroliagi, M. (2001). Rh (II)-binaphthol phosphate catalysts in the enantioselective intramolecular oxonium ylide formation–[3, 2] sigmatropic rearrangement of α-diazo-β-ketoesters. *Tetrahedron: Asymmetry*, *12*(6), 877-881.
49. Hoffmann, P., Held, C., Maskow, T., & Sadowski, G. (2014). A thermodynamic investigation of the glucose-6-phosphate isomerization. *Biophysical Chemistry*, *195*, 22-31.

50. Issitt, P. D., & Anstee, D. J. (1975). *Applied Blood Group Serology* (Vol. 24) pp,78-119

51. Jin, E. S., Sherry, A. D., & Malloy, C. R. (2015). Lactate contributes to glyceroneogenesis and glyconeogenesis in skeletal muscle by reversal of pyruvate kinase. *Journal of Biological Chemistry*, *290*(51), 30486-30497.

52. Jolley, K. A., Feil, E. J., Chan, M. S., & Maiden, M. C. J. (2001). Sequence type analysis and recombinational tests (START). *Bioinformatics*, *17*(12), 1230-1231.

53. Kane, J. P., Pullinger, C. R., Goldfine, I. D., & Malloy, M. J. (2021). Dyslipidemia and diabetes mellitus: Role of lipoprotein species and interrelated pathways of lipid metabolism in diabetes mellitus. *Current Opinion in Pharmacology*, *61*, 2

54. Kang, J. G., Park, S. Y., Ji, S., Jang, I., Park, S., Kim, H. S., ... & Cho, J. W. (2009). O-GlcNAc protein modification in cancer cells increases in response to glucose deprivation through glycogen degradation. *Journal of Biological Chemistry*, *284*(50), 34777-34784.

55. Kelley, W. N., Rosenbloom, F. M., Henderson, J. F., & Seegmiller, J. E. (1967). A specific enzyme defect in gout associated with overproduction of uric acid. *Proceedings of the National Academy of Sciences*, *57*(6), 1735-1739.

56. Kobe, B., & Kajava, A. V. (2001). The leucine-rich repeat as a protein recognition motif. *Current Opinion in Structural Biology*, *11*(6), 725-732.

57. Krebs, H. A. (1954). The tricarboxylic acid cycle. In *Chemical pathways of metabolism* vol.,1, Walsh, K., & Koshland, D. E. (1984). Determination of flux through the branch point of two metabolic cycles. The tricarboxylic acid cycle and the glyoxylate shunt. *Journal of Biological Chemistry*, *259*(15), 9646-9654. (pp. 109-171)

58. Krebs, H. A. (1970). The history of the tricarboxylic acid cycle. *Perspectives in biology and medicine*, *14*(1), 154-172. Krebs, H. A. (1970). The history of the tricarboxylic acid cycle. *Perspectives in Biology and Medicine*, *14*(1), 154-172.

59. Kumler, W. D., & Eiler, J. J. (1943). The acid strength of mono and diesters of phosphoric acid. The n-alkyl esters from methyl to butyl, the esters of biological importance, and the natural guanidine phosphoric acids. *Journal of the American Chemical Society*, *65*(12), 2355-2361.

60. Kunishima, F., Yotsumoto, F., & Kono, Y. (2007). A case of ductal carcinoma with squamous differentiation in malignant phyllodes tumor. *Breast Cancer*, *14*(3), 327-332.

61. Kuo, S. C., & Lampen, J. O. (1974). Tunicamycin—an inhibitor of yeast glycoprotein synthesis. *Biochemical and Biophysical Research Communications*, *58*(1), 287-295.

62. Kuzuyama, T., Takahashi, S., Takagi, M., & Seto, H. (2000). Characterization of 1-deoxy-D-xylulose 5-phosphate reductoisomerase, an enzyme involved in isopentenyl diphosphate biosynthesis, and identification of its catalytic amino acid residues. *Journal of Biological Chemistry*, *275*(26), 19928-19932.

63. Lands, W. E. (1965). Lipid metabolism. *Annual Review of Biochemistry*, *34*(1), 313-346.

64. Leonard, J. V., & Morris, A. A. M. (2002, February). Urea cycle disorders. In *Seminars in neonatology* (Vol. 7, No. 1, pp. 27-35).

65. Liu, J. Y., Chen, C. H., Lin, C. H., Tsai, H. F., Chen, C. H., & Kamogawa, M. (2011). Ionospheric disturbances triggered by the 11 March 2011 M9. 0 Tohoku earthquake. *Journal of Geophysical Research: Space Physics*, *116*(A6).
66. Lösel, R., & Wehling, M. (2003). Nongenomic actions of steroid hormones. *Nature Reviews Molecular cell biology*, *4*(1), 46-55.
67. Maan, M., Peters, J. M., Dutta, M., & Patterson, A. D. (2018). Lipid metabolism and lipophagy in cancer. *Biochemical and Biophysical Research Communications*, *504*(3), 582-589.
68. Mailloux, R. J., Beriault, R., Lemire, J., Singh, R., Chenier, D. R., Hamel, R. D., & Appanna, V. D. (2007). The tricarboxylic acid cycle, an ancient metabolic network with a novel twist. *PloS One*, *2*(8), e690.
69. Marcucci, F., & Rumio, C. (2023). On the Role of Glycolysis in Early Tumorigenesis—Permissive and Executioner Effects. *Cells*, *12*(8), 1124.
70. Markowitz, J. (1925). Glyconeogenesis. *American Journal of Physiology-Legacy Content*, *74*(1), 22-35.
71. Matsunaga, T., Suzuki, T., Kobayashi, S., Yamaguchi, H., Orba, Y., ... & Sawa, H. (2013). Gold nanoparticles as a vaccine platform: influence of size and shape on immunological responses *in vitro* and *in vivo*. *ACS Nano*, *7*(5), 3926-3938.
72. Matthews, D. E. (2007). An overview of phenylalanine and tyrosine kinetics in humans. *The Journal of Nutrition*, *137*(6), 1549S-1555S.
73. Merrick, W. C., & Nyborg, J. (2000). The protein biosynthesis elongation cycle. *Cold Spring Harbor Monograph Series*, *39*, 89-126.
74. Miflin, B. J., & Lea, P. J. (1977). Amino acid metabolism. *Annual Review of Plant Physiology*, *28*(1), 299-329.
75. Moja, E. A., Lucini, V., Benedetti, F., & Lucca, A. (1996). Decrease in plasma phenylalanine and tyrosine after phenylalanine-tyrosine free amino acid solutions in man. *Life Sciences*, *58*(26), 2389-2395.
76. Moriwaki, Y., Yamamoto, T., & Higashino, K. (1999). Enzymes involved in purine metabolism-a review of histochemical localization and functional implications. *Histology and histopathology*, *14*(4), 1321-1340.
77. Morot-Gaudry, J. F., Job, D., & Lea, P. J. (2001). Amino acid metabolism. *Plant Nitrogen*, 167-211.
78. Morris Jr, S. M. (2002). Regulation of enzymes of the urea cycle and arginine metabolism. *Annual Review of Nutrition*, *22*(1), 87-105.
79. Muhammad, M. R., Devi, G. P., & Selvakumar, R. (2013). A computational model for monitoring glycolysis process in cancer cells. *Journal of Bioinformatics and Intelligent Control*, *2*(4), 300-304.
80. Muronetz, V. I., Melnikova, A. K., Saso, L., & Schmalhausen, E. V. (2020). Influence of oxidative stress on catalytic and non-glycolytic functions of glyceraldehyde-3-phosphate dehydrogenase. *Current Medicinal Chemistry*, *27*(13), 2040-2058.

81. Nguyen, P., Leray, V., Diez, M., Serisier, S., Bloc'h, J. L., Siliart, B., & Dumon, H. (2008). Liver lipid metabolism. *Journal of Animal Physiology and Animal Nutrition*, *92*(3), 272-283.

82. Patra, K. C., & Hay, N. (2014). The pentose phosphate pathway and cancer. *Trends in biochemical sciences*, *39*(8), 347-354.

83. Picard, D. (2006). Chaperoning steroid hormone action. *Trends in Endocrinology & Metabolism*, *17*(6), 229-235.

84. Pilkis, S. J., El-Maghrabi, M. R., McGrane, M., Pilkis, J., Fox, E., & Claus, T. H. (1982). Fructose 2, 6-bisphosphate: a mediator of hormone action at the fructose 6-phosphate/fructose 1, 6-bisphosphate substrate cycle. *Molecular and Cellular Endocrinology*, *25*(3), 245-266.

85. Preiss, J. (1984). Bacterial glycogen synthesis and its regulation. *Annual Review of Microbiology*, *38*(1), 419-458.

86. Rae, C. D. (2014). A guide to the metabolic pathways and function of metabolites observed in human brain 1 H magnetic resonance spectra. *Neurochemical Research*, Vol,1 *39*, 1-36.

87. Ramasamy, I. (2014). Recent advances in physiological lipoprotein metabolism. *Clinical Chemistry and Laboratory Medicine (CCLM)*, *52*(12), 1695-1727.

88. Raviglione, M. C., & Uplekar, M. W. (2006). WHO's new Stop TB Strategy. *The Lancet*, *367*(9514), 952-955.

89. Richard, J. P. (1984). Acid-base catalysis of the elimination and isomerization reactions of triose phosphates. *Journal of the American Chemical Society*, *106*(17), 4926-4936.

90. Rocha, M., Licausi, F., Araujo, W. L., Nunes-Nesi, A., Sodek, L., Fernie, A. R., & Van Dongen, J. T. (2010). Glycolysis and the tricarboxylic acid cycle are linked by alanine aminotransferase during hypoxia induced by waterlogging of Lotus japonicus. *Plant Physiology*, *152*(3), 1501-1513.

91. Ryan, C. O. L. L. E. E. N., & Radziuk, J. E. R. R. Y. (1994). Muscle glyconeogenesis during recovery from a prolonged swim in rats. *American Journal of Physiology-Endocrinology and Metabolism*, *267*(2), E210-E218.

92. Sakakibara, R., Kato, M., Okamura, N., Nakagawa, T., Komada, Y., Tominaga, N., ... & Fukasawa, M. (1997). Characterization of a human placental fructose-6-phosphate, 2-kinase/fructose-2, 6-bisphosphatase. *The Journal of Biochemistry*, *122*(1), 122-128.

93. Scagliola, A., Mainini, F., & Cardaci, S. (2020). The tricarboxylic acid cycle at the crossroad between cancer and immunity. *Antioxidants & Redox Signaling*, *32*(12), 834-852.

94. Sears, P., & Wong, C. H. (1998). Enzyme action in glycoprotein synthesis. *Cellular and Molecular Life Sciences CMLS*, *54*, 223-252.

95. Seo, K. H., Ok, J., Son, J. H., & Cha, D. H. (2013). Assessing future changes in the East Asian summer monsoon using CMIP5 coupled models. *Journal of Climate*, *26*(19), 7662-7675.

96. Shore, L. S., & Shemesh, M. (2003). Naturally produced steroid hormones and their release into the environment. *Pure and Applied Chemistry*, *75*(11-12), 1859-1871.
97. Siiteri, P. K., Murai, J. T., Raymoure, W. J., Kuhn, R. W., Hammond, G. L., & Nisker, J. A. (1982, January). The serum transport of steroid hormones. In *Proceedings of the 1981 Laurentian Hormone Conference* vol,1 (pp. 457-510).
98. Smith, A. C., & Robinson, A. J. (2011). A metabolic model of the mitochondrion and its use in modelling diseases of the tricarboxylic acid cycle. *BMC systems biology*, *5*(1), 1-13.
99. Soine, P. J., Blanke, R. V., Guzelian, P. S., & Schwartz, C. C. (1982). Preferential binding of chlordecone to the protein and high density lipoprotein fractions of plasma from humans and other species. *Journal of Toxicology and Environmental Health, Part A Current Issues*, *9*(1), 107-118.
100. Sozio, M., & Crabb, D. W. (2008). Alcohol and lipid metabolism. *American Journal of Physiology-Endocrinology and Metabolism*, *295*(1), E10-E16.
101. Springer, G. F., & Horton, R. E. (1969). Blood group isoantibody stimulation in man by feeding blood group-active bacteria. *The Journal of Clinical Investigation*, *48*(7), 1280-1291.
102. Stalmans, W., Bollen, M., & Mvumbi, L. (1987). Control of glycogen synthesis in health and disease. *Diabetes/metabolism Reviews*, *3*(1), 127-161.
103. Stumpf, P. K. (1960). Lipid metabolism. *Annual Review of Biochemistry*, *29*(1), 261-294.
104. Taki, K., Noda, S., Hayashi, Y., Tsuchihashi, H., Ishii, A., & Zaitsu, K. (2020). A preliminary study of rapid-fire high-throughput metabolite analysis using nano-flow injection/Q-TOFMS. *Analytical and Bioanalytical Chemistry*, *412*, 4127-4134.
105. Tessari, P., Vettore, M., Millioni, R., Puricelli, L., & Orlando, R. (2010). Effect of liver cirrhosis on phenylalanine and tyrosine metabolism. *Current Opinion in Clinical Nutrition & Metabolic Care*, *13*(1), 81-86.
106. Toth, M. J., & Tchernof, A. (2000). Lipid metabolism in the elderly. *European journal of clinical nutrition*, *54*(3), S121-S125.
107. Tovar-Méndez, A., Mújica-Jiménez, C., & Muñoz-Clares, R. A. (1997). Desensitization to glucose 6-phosphate of phosphoenolpyruvate carboxylase from maize leaves by pyridoxal 5′-phosphate. *Biochimica et Biophysica Acta (BBA)-Protein Structure and Molecular Enzymology*, *1337*(2), 207-216.
108. Tsutsumi, K., Hagi, A., & Inoue, Y. (2001). The relationship between plasma high density lipoprotein cholesterol levels and cholesteryl ester transfer protein activity in six species of healthy experimental animals. *Biological and pharmaceutical bulletin*, *24*(5), 579-581.
109. Vaseghi, S., Baumeister, A., Rizzi, M., & Reuss, M. (1999). *In vivo*Dynamics of the Pentose Phosphate Pathway in *Saccharomyces cerevisiae*. *Metabolic engineering*, *1*(2), 128-140.

110. Wakil, S. J. (1962). Lipid metabolism. *Annual Review of Biochemistry*, *31*(1), 369-406.

111. Wang, T., & Himoe, A. (1974). Kinetics of the rabbit muscle enolase-catalyzed dehydration of 2-phosphoglycerate: fluoride and phosphate inhibition. *Journal of Biological Chemistry*, *249*(12), 3895-3902.

112. Wehling, M. (1997). Specific, nongenomic actions of steroid hormones. *Annual Review of Physiology*, *59*(1), 365-393.]

113. Williams, C. M. (2004). Lipid metabolism in women. *Proceedings of the Nutrition Society*, *63*(1), 153-160.

114. Wu, G. (2009). Amino acids: metabolism, functions, and nutrition. *Amino acids*, *37*, 1-17.

115. Wu, Q., Dohnal, V., Huang, L., Kuča, K., & Yuan, Z. (2010). Metabolic pathways of trichothecenes. *Drug Metabolism Reviews*, Vol,1 *42*(2), 250-267.

116. Xu, L., Zhou, L., & Li, P. (2012). CIDE proteins and lipid metabolism. *Arteriosclerosis, Thrombosis, and Vascular Biology*, *32*(5), 1094-1098.

117. Yang, S. H., Lim, J. S., Khan, M. A., Kim, B. S., Choi, D. Y., Lee, E. Y., & Ahn, H. K. (2015). High-throughput nucleotide sequence analysis of diverse bacterial communities in leachates of decomposing pig carcasses. *Genetics and Molecular Biology*, *38*, 373-380.

118. Yang, Y. Y., Ficht, S., Brik, A., & Wong, C. H. (2007). Sugar-assisted ligation in glycoprotein synthesis. *Journal of the American Chemical Society*, *129*(24), 7690-7701.

119. Yu, D., Richardson, N. E., Green, C. L., Spicer, A. B., Murphy, M. E., Flores, V., ... & Lamming, D. W. (2021). The adverse metabolic effects of branched-chain amino acids are mediated by isoleucine and valine. *Cell Metabolism*, *33*(5), 905-922.

120. Zimmermann, H. (1996). Extracellular purine metabolism. *Drug Development Research*, *39*(3-4), 337-352.

CHAPTER 8

SUSTAINABLE CHEMICAL AND BIOCHEMICAL TECHNOLOGIES

UNIT INTRODUCTION

This chapter provides an in-depth discussion of the development trends in current chemical technologies as well as the fundamental modifications in technology that have been made to promote environmental sustainability. It is also possible to identify the fundamental approaches of sustainable development as well as prospective alternatives. A number of forecasts and points of view are presented in relation to the likely progression of the expansion of cutting-edge and environmentally friendly chemical technologies.

The historical periods of agricultural, industrial, and modern scientific-technical growth have all played a significant role in determining the course that human civilization will take. At the moment, we are living in an age characterized by cutting-edge information technology and environmentally responsible practices. The culture is undergoing a transition into a brand-new era of environmentally responsible development (Hongbo et al., 2010; Rahmani et al., 2012; Suraweera et al., 2014). When people hear the word “sustainable development,” they often think of a development that is balanced or stable. Words such as trway rozwj, zrwnowaz ‘ony rozwj, nachhaltige Entwicklung, and development durable all refer to the same concept, but represent diverse understandings of what it means. This is a reflection of the fact that sustainable development is understood differently in different languages. Since it was first conceived, the idea of sustainable development has undergone numerous iterations. Over the course of the past two decades, the ultimate goals, characteristics, and definitions of sustainable development have undergone substantial transformations (Fukui et al., 1994; Nishikawa et al., 2001; Levenspiel, 2005; Jena, 2009). The idea of maintaining a healthy environment for future generations has been advanced by a number of scientists and environmentalists (Meadows et al., 1972; Keeble, 1988; Butlin, 1989). At a symposium hosted by the United Nations Conference on Environment and Development, a form with greater potential was presented and discussed (Houghton et al., 1992; Grubb, 1993; Elkington, 1994). A great number of authors have presented in-depth discussions on the modern concepts of sustainable living (Borowiecki et al., 1998; Taniewski, 1999; 2012). The term "sustainable development" refers to a human idea that is geared on improving the standard of living and general well-being of people. The advancements in sustainable development are being made in order to meet the needs of future generations despite the difficult conditions that will be present, namely a scarcity of resources and an abundance of industrial operations. A progressive worldview brings together issues of social justice, protecting the environment, and expanding the economy (Taniewski, 2004; 2005; 2006; 2008). It is widely held that sustainable development may be gradually put into practice, despite the fact that there are many challenges and obstacles to overcome. The traditional models of civilization development that maximized manufacturing and new models of development that focus on sustainable development and its implementation tactics will replace profits. At the present time, sustainability is not an option but rather a vital imperative for the human race (Ismagilov et al., 2011; Bednarek-Gejo et al., 2012).

TIPS

Be aware that the development of sustainable technologies is critical to addressing many of the environmental challenges we face today. Support companies and organizations that are working to develop more sustainable alternatives to traditional chemical and biochemical processes. Look for ways to reduce your own environmental impact by making small changes in your daily life, such as using reusable bags and water bottles.

For the innovative and environmentally friendly solutions to be put into action, significant adjustments need to be made to the structure of the economy, consumption and production patterns, technologies, organizations and institutions, environmental and industrial ethics, and other areas. The introduction of new technological advancements plays an important part and acts as a crucial driver in the changing of fundamental models of development (Degnan, 2000; 2003; Catani et al., 2002; Marcilly, 2003).

At this time, the most successful chemical industries in the world have realized the significance of implementing sustainable growth methods (Arpentinier et al., 2005; Cavani et al., 2005; Muda, 2014). A number of specialized programs, including Responsible

Care, Product Stewardship, Life Cycle Assessment, and cradle-to-grave, are about to be introduced in an effort to promote environmental preservation and the development of environmentally friendly chemical technologies (Bozell, 2001; Wilson & Schwarzman, 2009). It is a widely held conviction that the gradual adoption of sustainable models in the chemical industries, along with ongoing efforts to enhance those models, will inevitably lead to significant advances in technology. For example, Chum and Overend (2001) anticipated that by the year 2020, there will be a 90% reduction in feedstock losses owing to wastes and by-products. In addition to that, it is anticipated that the utilization of carbon compounds and renewable energy resources will rise by 20% and 13%, respectively (Anastas & Kirchhoff, 2002; Trost, 2002; Anastas & Zimmerman, 2003)

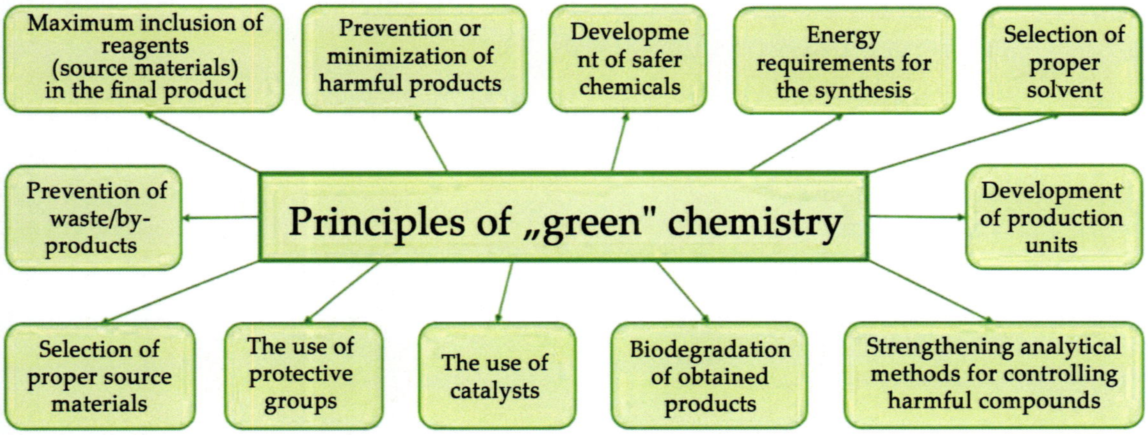

Figure 8.1. Illustration of the principles of green chemistry and its applications (Source: Soltys et al., Creative Commons License).

The following tendencies in contemporary environmentally friendly chemical technologies have been the subject of a large number of publications, including research articles, general reviews, and presentations (Metzger, 2006; Guillena et al., 2007; Chen et al., 2008; Biermann et al., 2011). The topics of green chemistry, carbon chemistry, renewable raw materials, and methane have been the subject of some significant reviews that have been published in the field of sustainable development (Lunsford, 1995; 2000; Xu et al., 2003; Holmen, 2009).

Additional aspects of sustainability include the development of clean fuels, syngas, trends in catalysis, nanotechnology, miniaturization, and separation technologies, as well as innovations for sustainability (Kioes & Liebner, 2004; Koempel et al., 2005; Liebner & Koempel, 2005; Pushparaj et al., 2013). Some publications include in-depth ideas for sustainable living that are a synthesis of traditional wisdom and cutting-edge research (Taniewski, 2002; Aramendía et al., 2004; Skutil & Taniewski, 2006; Nouralishahi et al., 2008). This chapter's focus will be on a comprehensive examination of the many contemporary chemical technologies. In the following sections of this chapter, we will also discuss the fundamentals of environmentally friendly technological engineering.

Learning Objectives

At the end of this chapter, the reader will be able to understand

- The chemical technologies in development
- The importance of technological processes
- The concept of biotechnology
- The process of chemical nanotechnology
- The recent developments in green technology
- The uses of alternative chemicals as in chemical industry

Key Terms

- Biotechnology
- Nanotechnology
- Feedstock
- Gas-To-Liquids (GTL)
- Sheldon's E-factor
- Green Technology

8.1. CHEMICAL TECHNOLOGIES IN DEVELOPMENT

The following is a list of the primary areas that are receiving the most attention from worldwide trends toward sustainable (green) chemical technologies:

- A reduction in the number of raw materials that are used, which is made possible by an improvement in selectivity; this improvement is the result of a wide variety of options, including acceptable processes, optimal operating parameters, feedstock, recirculation of by-products, highly selective catalysts, etc. (Van-Hove,2006; Gawlitza et al., 2007; Erdély et al., 2009; Wang et al., 2009)
- The reduction in power consumption is made possible by the increase in selectivity brought about by a variety of options, including the development of exothermic & endothermic operations & processes, the gradual replacement of extremely energy-intensive cryogenic separation processes with adsorption and membranes, the advancement of the heat exchanger and reactor technology, and innovations in steering and process control, etc. (Charpentier, 2005; Sell et al., 2014; Zhang et al., 2014; Francisco et al., 2015; Shanmugam & Donaldson, 2015)
- Raising the bar for environmental safety by adhering to environmental norms, which include reducing emission levels, disposing of waste and sewage properly, developing zero-waste or low-waste technologies, enhancing the safety of processes, etc.
- Industrial waste management techniques that include the safe processing, disposal, destruction, recycling, and combustion of industrial wastes.

The above-mentioned behaviors and dispositions are consistent with the core ethical values. To control the many industrial processes used today, ethics plays a crucial role. The continual danger and obligations for chemical processes, goods, and their safety are related to the ethical concerns. Therefore, it is crucial to uphold moral principles in order to assure one's own development, the elimination of risks and threats, open communication with others, and reliable information practices.

For the advancement of well-known technologies as well as the creation of unique, environmentally friendly, and sustainable chemical

KEYWORD

Nanotechnology refers to the branch of science and engineering devoted to designing, producing, and using structures, devices, and systems by manipulating atoms and molecules at nanoscale, i.e. having one or more dimensions of the order of 100 nanometres (100 millionth of a millimetre) or less.

technologies, a variety of tools are available. The following list of fundamental instruments is provided in order to attain technological sustainability:

- Choosing the best primary resources to replace existing ones
- Choosing the most appropriate alternative feedstocks
- Selecting the best operating procedures and processes
- The best energy sources and energy management systems should be chosen
- Choosing the best catalysts possible
- Choosing the best reactors and equipment systems
- Choosing the best techniques for managing sewage, waste, emissions, and by-products
- Choosing the best and most secure items

KEYWORD

Heat exchanger is a system used to transfer heat between a source and a working fluid. Heat exchangers are used in both cooling and heating processes.

These fundamental instruments depend on one another. Typically, the choice chosen in one area influences the choice in the other. It is common for synthesis processes and routes to vary in response to changes in feedstocks and main resources. Continuous work will always be needed to transform common technology into sustainable, environmentally benign, technically ideal, and economically viable technologies. The most sustainable technologies among those that are currently available can be chosen using a variety of assessment methodologies, such as the best available techniques (BAT).

8.2. PRIMARY RESOURCES AND FEED

Refinery gas streams, crude oil fractions, natural gasoline, natural gas, LPG, coal liquids, biomass, coal gases, minerals, and atmospheric CO_2 are the basic resources that can be used as the starting materials for the synthesis of intermediates (such as feedstocks, synthetic fuels, or chemical compounds).

It is firmly believed that petrochemical-based refinery gas streams, crude oil fractions, natural gas, natural gasoline, and LPG will continue to be the main bulk products of commercial inorganic manufacturing (hydrogen, nitrogen, ammonia, fertilizers, etc.) and a crucial component of commercial organic manufacturing (polymers, synthetic rubber, synthetic fibers, detergents, etc.). On the other hand, the slow increase in oil prices will spur the development of new forms of alternative energy. The processing and use of crude oil's heavy and extremely heavy byproducts are the responsibility of the crude oil fraction group. These anticipated trends will result from an increased requirement to employ heavy fractions and residues, as well as from the creation of heavy and super-heavy crude oil products in substitute of progressively depleting conventional oils.

The relative contribution of methane (natural gas) to primary hydrocarbon resources is predicted to rise at the expense of oil in a variety of uses, including petrochemical , transportable liquid fuels, and energetics. (gas-to-liquids, GTL). Large natural gas reserves (higher yearly production ratio/proved reserves) and recent advancements in CH_4 chemical technology are the primary factors that have made this significant change conceivable. Abundant resources of flared gas(Flared gas, also known as flare gas, is natural gas that is burned off at oil and gas production facilities through a process called flaring), sparsely used natural gas in remote areas, and even sizable reserves of unexplored methane hydrates will spur their general processing and utilization.

In the late 18th and early 19th centuries, coal (as a primary raw material) made a significant contribution to the energy sector. However, in favor of natural gas and petroleum, the contribution of coal and its byproducts (coal tar and coke-oven gas) dramatically decreased. Currently, coal's use is restricted to the manufacture of aromatics and hydrogen from natural gas and oil on a small scale. Since the 1950s, South Africa has engaged in large-scale Fischer-Tropsch synthesis and coal gasification, which are both uncommon occurrences brought on by particular conditions. However, the introduction of gas-based synthesis routes starting in the 1990s caused a significant depreciation of the coal-based synthesis pathways (Mossel Bay). Despite the evidence, it is reasonable to believe that the commercial coal industry will gradually recover in the years to come, especially in the areas of new gasification techniques leading to advanced coal-based materials (composites, fibers, sorbents, and carbon nanotubes), coalbed methane processing, and SNG (Substitute natural gas). These new fields have lately experienced significant advancement. However, it is extremely improbable that carbonization (coal degassing) processes such as coal hydro-liquefaction or coking will resume.

Biofuels, fine chemicals, or products with improved environmental biodegradability, such as polymers, are all produced through the chemical processing of renewables/biomass (fats, cellulose, vegetable oils, agricultural residues, municipal wastes, etc.). As a result, these procedures ought to be taken into account as

KEYWORD

Best Available Techniques (BAT) are advanced and proven techniques for the prevention and control of industrial emissions and the wider environmental impact caused by industrial installations, which are developed at a scale that enables implementation under economically and technically viable conditions.

alternatives for processing energy products. Prospective process considerations must be made in order to produce chemicals in large quantities from these special resources.

8.3. FEEDSTOCK

KEYWORD

Natural gas is an odorless, gaseous mixture of hydrocarbons—predominantly made up of methane (CH4). It accounts for about 30% of the energy used in the United States.

The best raw materials for feedstock at the moment are those that can be most selectively transformed into desirable products. For the production of higher and lower olefins, di-olefins, surfactant intermediates, alkyl aromatics, oxygen compounds, heterocyclic compounds, etc., a variety of alternative feedstock raw materials can be used (Noble & Agrawal, 2005; Broering et al., 2006). There are often a lot of different alternative feedstocks used in commercial organic synthesis, which leads to a lot of different processing pathways that can lead to isoprene. This chapter has already covered a few substitute feedstocks and alternative processing methods used in the petrochemical sector (Taniewski, 2002; Koros et al., 2004). As mentioned above, there are more and more applications for renewable (novel) feedstocks. The sections that follow explain the signs of feedstock transformation processes. Below are a few illustrations of alternative feedstocks and primary resources employed in the synthesis of the bulk intermediates:

- Ethylene (oxidative dehydrogenation or dehydrogenation of ethane and oxidative coupling of methane; steam cracking of LPG, gasoline, ethane, heavy, and middle oil fractions)
- Propylene (oxidative dehydrogenation or dehydrogenation of propane, separation from FCC gases, steam cracking of liquid hydrocarbon fractions or LPG, and conversion of methanol)
- Dehydrogenation of n-butenes and n-butane, as well as steam cracking of liquid hydrocarbon fractions, produce butadiene.
- Steam cracking of liquid hydrocarbon fractions, catalytic reformation of gasoline, hydrotreatment of the pyro-gasoline fraction, hydrodealkylation of toluene, and toluene disproportionation are some methods for producing benzoene.
- Toluene (catalytic reforming of gasoline and steam cracking of liquid hydrocarbon fractions)
- Steam cracking of liquid hydrocarbon fractions, catalytic reforming of gasoline, and alkylation or disproportionation of toluene are all examples of producing xylenes.

- Synthesis gas, SNG (gasification of coal, processing of biomass, reforming of heavy oil fractions, and partial oxidation or steam reforming of natural gas)
- Hydrogen (reforming of oil waste or natural gas, gasification of biomass or coal, reforming of renewable liquid fuel (bio-oils, bioethanol, etc.), reforming of DME, ethanol, methanol, reforming of hydrocarbon fuels for cells, water decomposition by high-temperature splitting process, catalytic breakdown of natural gas, electrolysis of water using renewable sources of molecular or electrical energy, photolytic splitting of water using photo electrolytic or photobiological methods, etc. (Dijkema et al., 2006; Davood et al., 2016;).

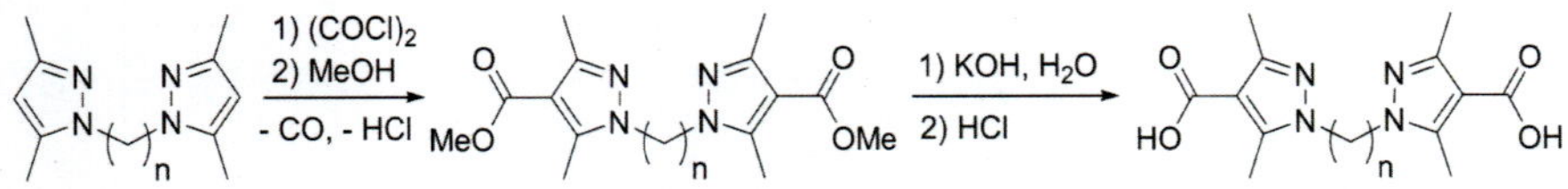

4a (n = 4) **5a** (n = 5) **6a** (n = 6)

4c (n = 4), 27% **5c** (n = 5), 34% **6c** (n = 6), 30%

4b (n = 1), 82% **5b** (n = 2), 78% **6b** (n = 3), 80%

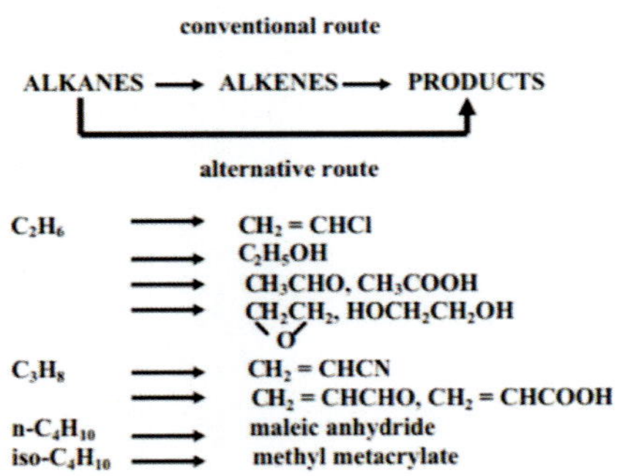

Figure 8.2. Schematic of Alkane derivatives (Source: Burlutskiy et al., Creative Commons License).

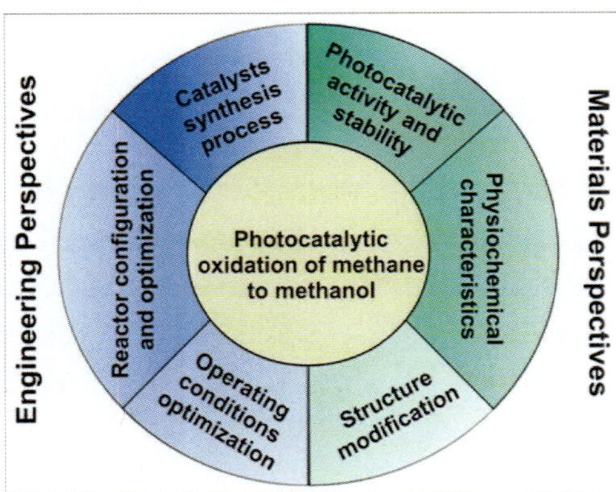

Figure 8.3. Schematic of production and R&D or methane derivatives (Source: Yuniar et al., Creative Commons License).

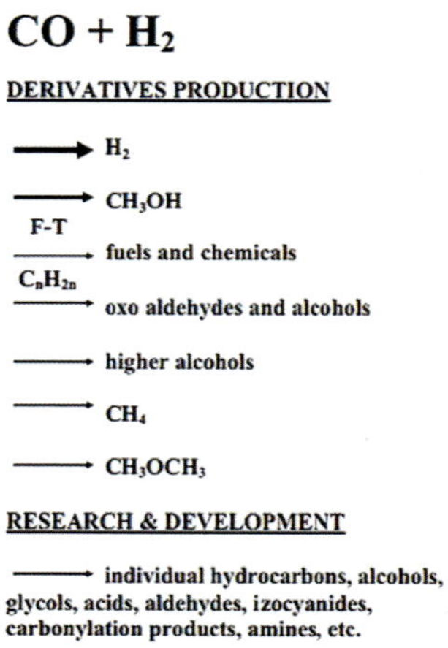

Figure 8.4. Production and R&D of synthesis gas derivatives, i.e., F-T – Fischer-Tropsch synthesis (Source: OnlineLibrary, Creative Commons License).

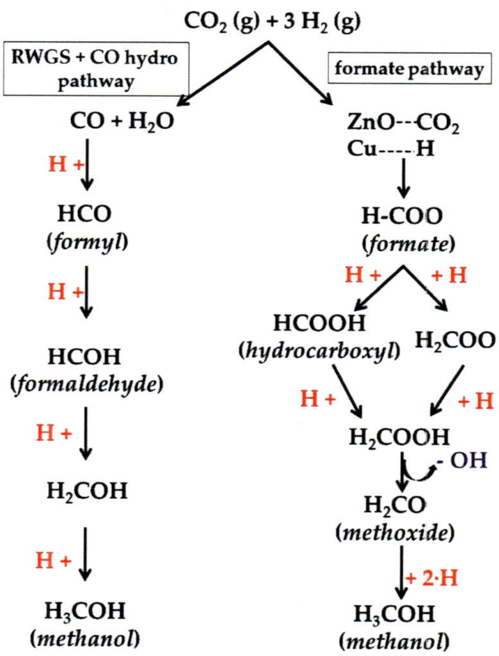

Figure 8.5. Pathways of methanol synthesis from CO_2 *hydrogenation (Source: Guil-Lopez et al., Creative Commons License).*

The choice of a specific feedstock material may occasionally lead to a desired reduction in the number of synthesis steps, avoiding the production of intermediate compounds (for example, the production of alkene derivatives from alkanes in place of alkenes, the synthesis of C1 derivatives from methane in place of syngas, and vice versa). Over the past two decades, it has been noted that certain alternative methods for producing alkanes-based derivatives of alkenes have been used. These straightforward yet reliable synthesis techniques will compete with traditional alkene-based techniques (Lepoutre, 2008; White, 2009).

The possibility of using the same feedstock in the same industry for the synthesis of different products, or the requirement to choose the best production profile, are examples that highlight yet another important issue with the choice of feedstock (Lorenz & Eck, 2005; Nouralishahi et al., 2008).

KEYWORD

Feedstocks are the raw ingredients for composting. They are organic materials, usually solid, and usually in an active state of decomposition.

8.4. TECHNOLOGICAL PROCESSES

Petrochemical, carbo-chemical, refining, biochemical, environmental, mineral-based (including hydrometallurgy and ceramics), and speciality chemical processes utilizing membranes, fuel cells, and ionic liquids are some key technical processes that can be classed. The tendency to produce highly selected zero-waste or low-waste technology often serves as the foundation for the pursuit of the finest technical process. It frequently calls for choosing the optimum processing path from among numerous steps that result in the creation of the same products even from the same set of raw materials. There are cutting-edge synthetic technologies for producing transportable liquid fuels from natural gas via many (current/new) alternative methods (Lorenz & Zinke, 2005; Paula & Birrer, 2006).

8.4.1. Operating Conditions

Designating the type of process and its normal operating circumstances is crucial after selecting the optimum processing route. Low-pressure, high-pressure, liquid-phase, gas-phase, multiphase (mostly perspective in a few situations), catalytic, noncatalytic, or multi-catalytic processes are among the several types. The ideal operating conditions, on the other hand, are established in accordance with the process requirements for using supercritical fluids and other alternative materials.

8.4.2. Catalytic Technologies

Modern catalytic technologies can be used in specific chemical fields like petrochemicals and refining, among others. The possibility for the best selection and ongoing development is increased by the quick progress in the evaluation and use of the novel selective catalysts (regio- and stereoselective). Shape-selective zeolites, nanostructured, monolithic, mesoporous, blocks, metallocenes, and various one-centered catalysts, as well as heterogenized homogeneous, polyfunctional, bio, phase transfer, membrane, and other common catalysts, are some examples of common catalysts. (Sheldon & Dakka, 1994; Sheldon, 1996; Sheldon et al., 1998).

It is crucial to remember that in the 20th century, chemical technologies mainly emphasized on catalytic activities. The high selectivity of the catalysts, however, worries contemporary chemical technology more. This change in emphasis is understandable in light of the earlier technologies' tendency to have high catalytic activity as well as their higher selectivity under the present sustainability tendencies. The fact that this technology can lower the use of energy and raw materials gives the increase in process selectivity its crucial role. Additionally, the expenses associated with waste treatment and product separation are crucial in determining the function of catalytic selectivity.

KEYWORD

Liquid phase one of the three fundamental structural phases of matter in which the thermal mobility of molecules or atoms is comparable with the cohesiveness, having them connected but fluid in a mass.

On the other hand, several catalytic unit processes (such as partial oxidation, hydrogenation, and stereospecific polymerization) occur with subpar selectivity criteria. The use of catalytic technologies in the synthesis of green chemicals and medicinal products is yet in the future. Presently, the synthesis of fine chemicals and other related goods clearly favors noncatalytic pathways, resulting in unfavorably greater values of Sheldon's E-factor (cumulative waste mass/cumulative mass of the intended product), which is equal to 5100 (Anastas et al., 2000; Anastas & Kirchhoff, 2002).

The synthesis of all fine and specialty chemicals, including pharmaceuticals, intermediates, dyes, agrochemicals, pigments, fragrances, flavors, liquid crystals for displays, chemical compounds for the cosmetic and food industries, and electronic chemicals for the synthesis of chips, exhibits strong trends toward the development of safer and cleaner technologies (non-solvent, catalytic, etc.) (Anastas & Zimmerman, 2003; McDonough et al., 2003; Anastas, 2008).

8.5. BIOTECHNOLOGY

A lot of research is being done right now on the significance of industrial biotechnology as a replacement for conventional technologies based on minerals and fossil fuels for the chemical treatment of renewables and biomass (Anastas, 1998; Ballini, 2009; Sharma, 2010). Utilizing cutting-edge biotechnology, biofuels are synthesized by fermenting, trans-esterifying, or hydrolyzing renewable materials. (esters, alcohols, etc.). Furthermore, enzymatic manufactures of intermediates, fine chemicals, pharmaceuticals (such as vaccines, biomolecule synthesis, stereo-selective biocatalysts, and chiral technologies), and numerous biodegradable products are being carried out using renewable resources. Processing biomass or using biotechnological methods for chemical operations and bulk chemical synthesis show substantial but sluggish improvement (e.g., biodesulfurization of solids or liquids and enzymatic manufacture of 1, 3-propanediol from glucose molecules).

8.6. CHEMICAL NANOTECHNOLOGY

A subfield of nanotechnology, chemical nanotechnology emerged as a byproduct of the emergence of nanoscience from the general science of surfaces and thin films (Tang & Zhao, 2009; Soetaert & Vandamme, 2010). Nanomaterials (nanowires, nanofibers, nanotubes, etc.), catalysts, surfactants, sensors, membranes, coatings, polymers, etc. are all connected to this multidisciplinary, cutting-edge science. This multidisciplinary field of study focuses on the creation of unique and specialized products through the use of rapidly expanding alternative synthesis methods (Hopwood et al., 2005; Cheng & Gross, 2011).

8.7. CHEMISTRY AND ENVIRONMENT

The globe is currently dealing with climatic and environmental difficulties, which are widely acknowledged as being difficult issues. Sustainable development is crucial because environmental and green technologies are causing a decline in life quality. Technology has advanced significantly, especially with regard to clean air and water, which is depleting natural resources for life (Lélé, 1991; Sneddon et al., 2006). Their issues are seriously harming global social, environmental, and economic conditions. According to sustainable resources, renewable resources should be used wherever possible and non-renewable resources should be managed (e.g., recycled and decreased) to extend their useful lives for future generations (Irwin, 1995; Patil, 2013; 2014).

This chapter provides a review of the movement toward sustainable practices and green technologies in the chemical processing industry. A thorough examination of contemporary green technology is explored in terms of applications and understanding of sustainability, along with several case studies that show the

Remember

Biotechnology is the use of living organisms or their products to create new products or processes. This field has revolutionized medicine, agriculture, and many other industries. However, biotechnology also raises ethical and social questions about the use of genetic engineering and the manipulation of living organisms.

financial advantages of integrating green technologies from the perspective of chemical engineering. Green technologies use renewable resources to lessen waste, pollution, and emissions; reuse, recover, and recycle; rebalance the biosphere and echo-system; ease the strain on natural resources; and, in the end, help provide "ecologically sustainable development."

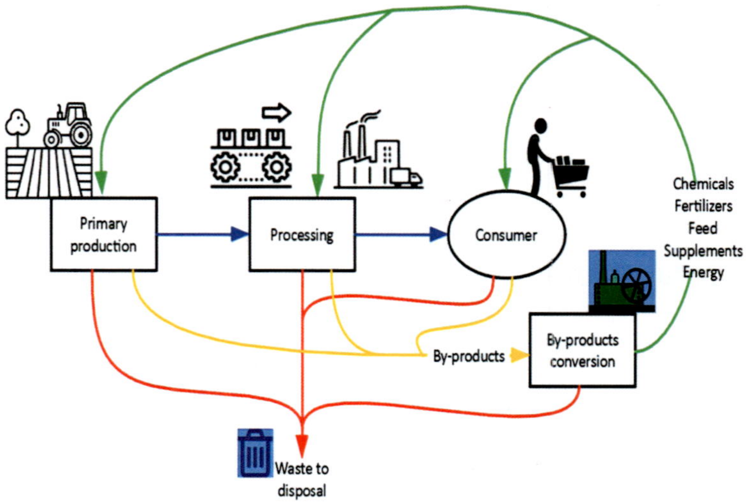

Figure 8.6. Illustration of future trends in green chemistry and its impact on environmental sustainability (Source: Ncube et al., Creative Commons License).

These green technologies are therefore affordable, practical, suitable for the country's climate, geographical, social, ecological, and economic requirements, and environmentally advanced (Dalton, 1998; Renn, 2008). The only way to accomplish this goal may be to develop new non-toxic, safe, and environmentally friendly products. (Pace, 2012). Therefore, CPI must promote sustainable development and assure increased adherence to health, safety, and environmental standards by investing in green technologies (Albrecht et al., 2006; Polshettiwar et al., 2009).

The rapid exploitation of natural resources and the economic system were once thought to be related to environmental issues. Horváth and Anastas (2007) and Anastas and Eghbali (2010) are two references to this. It required a long time to take into account the established relationships between the initial CPI design, the materials used (feedstocks), the energy used, the hazardous qualities of the products, and other factors that are involved in the production of the goods (recycling, lifecycle, etc.)

Environmentalists have become more interested in taking corrective action for the negative effects as a result of the growing development of some creative chemical goods and new chemical technologies in recent decades. (environmental pollution, monitoring, recycling, reduction of pollutants, etc.). It should be highlighted that the best method to reduce negative effects is to develop manufacturing processes, taking into account materials, use, energy, generation of secondary and atom economies, and at the conclusion of product life cycles, their recycling into new materials.

Green chemistry involved a more environmentally friendly synthetic approach, and it was an additional step in the industrial-scale production of chemicals with the motto "Think Green." For chemical engineers, selectivity has been more important than conversion. The study of alternative (solvent-free) media, reaction conditions, and alternative energy sources is known as "green technology." (Li & Trost, 2008). The development of chemical processes and products that eliminate or significantly minimize the production and use of hazardous compounds is the most fundamental strategy for preventing pollution. The necessity to create commodities and services that society depends on while also being environmentally benign is addressed in green chemistry. By implementing green chemistry concepts to chemical processes and products, it is possible to make plastics from plants that disintegrate, produce pharmaceuticals with less waste, and run chemical reactions in water rather than using conventional organic solvents.

KEYWORD

Phase transfer is the physical phenomenon where particles or molecules are transported between two immiscible solvents. In order to transfer particles from one solvent to another, the chemical properties must be altered in a way that allows for stable, low-energy interaction between particles and new solvent molecules.

8.8. SUSTAINABILITY CONCEPTS

Since the dawn of the twenty-first century, sustainable global development has come to be seen as the standard for both the preservation of the environment and the growth of the world economy. Environment, community, and economy are the three pillars of sustainable development. Therefore, in order to create a sustainable society, the next step must be followed.

- The natural step
- Eco-efficiency
- Pollution prevention
- Eco-effectiveness
- Design for environment
- Cradle-to-cradle design
- Industrial Ecology
- Environmental management
- Sustainable Management Systems

8.9. RECENT DEVELOPMENTS IN GREEN TECHNOLOGY

Scientists from all around the world are using their inventive and creative abilities to create new synthetic processes, reaction conditions, analytical tools, catalysts, etc. within the framework of modern green technology. Among them are:

The biomass is transformed into industrial fuels, chemicals, and animal feed using a continuous apparatus and process. Another method turns biomass, including sewage sludge, municipal solid waste, agricultural byproducts, tires, and plastic, into usable substances like acetic acid, ethanol, and hydrogen.

- A technique for semi-continuous *Taxus* genus plant culture for the industrial manufacture of taxol
- Fermentation-based synthesis of carboxylic acids.
- Through a method of partially oxidizing alcohol, such as esters or acids that convert methanol to aldehydes and ethers, employing a supercritical fluid mobile.
- By creating a fluoropolymer utilizing a technique involving supercritical carbon dioxide.

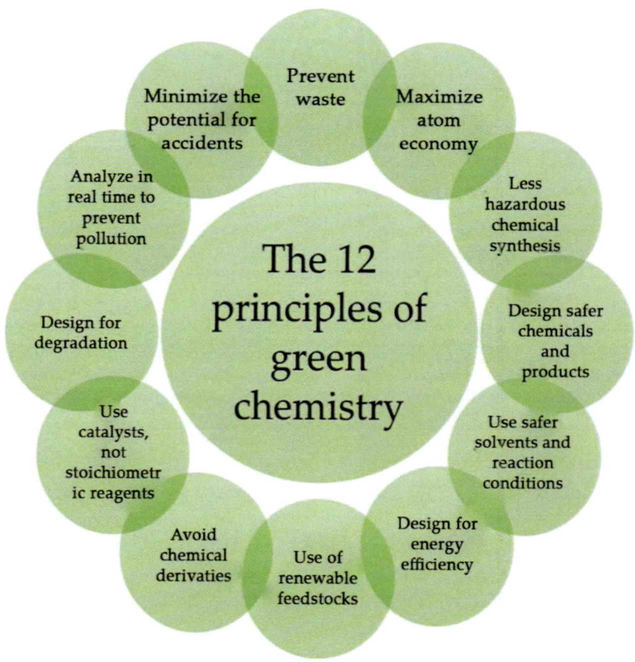

Figure 8.7. Schematic of how green chemistry is genuinely sustainable (Source: Trombino et al, Creative Commons License).

- The inexpensive production of ethyl lactate, a non-toxic solvent obtained from corn.

- A range of environmentally benign and user-friendly organic solvents, such as bioethanol.
- A cutting-edge, environmentally friendly approach has been used to recover zinc and ferrous chloride from pickle liquor.
- The need for non-ionic surfactants is rising. An inventive example derived from saccharide is alkyl glycoside. In shampoos, this substance can take the place of alkyl aryl sulphonate anionic surfactants.

8.10. USE OF ALTERNATIVE CHEMICALS IN CHEMICAL AND RESEARCH INDUSTRIES

Because of the petrochemical industry's dominance, the bulk of beginning materials and raw chemicals from not just the chemical industry but also from other significant industries have been known for the past 50 years.

KEYWORD

Biomass is renewable organic material that comes from plants and animals.

8.10.1. Renewable Raw Materials and Feedstocks

Green chemical technology must be converted into renewable feedstocks. The second most sought quality of the beginning materials is lower toxicity and environmental impact. The protection of the environment, workers' health, and safety comes first. Green chemistry (biogas, plant and animal waste, products from the fermentation of plant waste, etc.) suggests changes in the direction of biological raw materials.

8.10.2. Use of Oleochemicals as Modern Biological Materials

As oleochemical raw materials, oils and fats (from both plants and animals) might be thought of as a new source of chemical feedstocks. The market has some raw ingredients with numerous uses in lubricating oils, cosmetics, polymers, and other goods.

8.10.3. Light Aided Photochemical Reactions

Green technology places a strong emphasis on photochemical reactions in chemical processes. Visible and ultraviolet light can act as the primary catalyst in a variety of processes, in place of numerous harmful metals. According to several experts, photochemistry has a tremendous amount of potential for research applications and innovation in recent years.

8.10.4. Photocatalytic synthesis with Titanium dioxide (TiO2)

The use of TiO2 dioxide for photocatalytic processes at an industrial scale with visible light has received a lot of attention in recent years. The waste products are minimal, and the yields are significantly higher than in classical reactions.

KEYWORD

Polymer is any of a class of natural or synthetic substances composed of very large molecules, called macromolecules, which are multiples of simpler chemical units called monomers.

8.10.5. Decomposition of Toxic Chemicals and Photocatalytic Oxidation Waste

TiO2 and other metallic oxides (Fe_2+) can be employed in photocatalytic oxidations to break down waste and hazardous chemical substances. These decompositions can result in the production of neutral substances with low toxicity, particularly polychlorinated compounds, phenols, etc. Fe_2+/H_2O_2 is a combination that can be used to use light to aid in the decomposition of toxic industrial waste. (Fenton reagent)

8.10.6. Waste Biomass as Biomaterials Chemical Feedstock, and Biofuels

In the past ten years, there has been a notable progress in the utilization of biomass for the manufacture of diverse products. For years, it had been recognized that biomass from agricultural activities was wasted. The experts looked into the biomass's many facets and effectiveness. The value of fossil fuels has increased, and biomass is now seen as a major sustainability issue.

8.10.7. Degradation of Biomass for Biodiesel and Biogas

Biofuels are a widely accepted use of biomass, particularly the organic waste from landfills. Chemical and physical procedures can be employed to turn biomass into biodiesel.

8.10.8. Biotransformation and Biocatalysis in the Chemical Industry

Biocatalysis is one type of green technology that has a wide range of uses and is thought to be energy and environmentally friendly. Enzymes have been utilized for synthetic chemical pathways in the food and pharmaceutical sectors, with some success.

Did you Know?

The use of polyfluorinated phases, such as perfluorohexane, as a medium for organic synthesis has shown promising results in improving reaction efficiency and selectivity while reducing waste and toxicity.

8.10.9. Sequestration or Capture of Carbon Dioxide

Green chemistry is linked to the decrease of carbon dioxide in the chemical industries. A major environmental issue is the greenhouse effect and climate change brought on by CO2 emissions.

8.11. NOVEL FABRICATION TECHNOLOGIES FOR THE SYNTHESIS OF CHEMICAL COMPOUNDS

This section discusses some significant improvements in chemical manufacture under green chemical engineering and alternative techniques.

8.11.1. Ionic Liquids in Organic Synthesis

Ionic liquids have recently become an alternate solvent for chemical synthesis. Ionic melts, liquid electrolytes, fused salts, ionic glasses, or liquid salts are the names given to these compounds. Ionic liquids are strong solvents and electrically conductive fluids with a variety of uses. They are viewed as viable candidates for future advancement who can provide green credentials for their use and applications.

8.11.2. Organic Synthesis in Aqueous Medium

For many years, water was avoided as a solvent for synthetic organic chemistry and was instead treated as a medium. Water is regarded as an effective solvent for many synthetic processes. Water as a solvent is most interestingly used in Diels-Alder organic synthesis.

8.11.3. Organic Fabrication in Polyfluorinated Phase

Chemists use a two-phase solvent system known as polyfluorinated in these methods to dissolve catalysts with long aliphatic chains or hyperfluorinated alkyls. Reagents are dissolved by the organic solvent, which is insoluble in the hyperfluorinated phase. The reaction is accelerated by heating the mixture, producing good product yields.

KEYWORD

Biofuel is a fuel that is produced over a short time span from biomass, rather than by the very slow natural processes involved in the formation of fossil fuels, such as oil.

8.11.4. Supercritical Water and Supercritical Carbon Dioxide

Any liquid at pressure and temperature over the critical point is referred to as a supercritical fluid since there are no separate gas and liquid phases. Like a liquid, it can dissolve things, and like a gas, it can diffuse through solids. Additionally, around the critical point, a slight change in temperature or pressure causes a substantial change in density, enabling the "fine-tuning" of numerous supercritical fluid properties.

8.11.4.1 Use of Microwave Processes for Organic Chemical Synthesis

Microwave ovens are frequently used for cooking and warming meals. They have been employed in organic synthesis for a long time, and their success using "green" synthesis is well known. Numerous applications and research publications on the subject of high yield organic synthesis without solvent and with minimal energy needs currently exist.

8.11.5. The Use of Ultrasonic Waves for Chemical Synthesis

Modern green technologies use chemical reactions that are accelerated by ultrasonic and sonic waves and have a high yield. Sono-chemical reactions can be classified into three categories: heterogeneous Beyond the earlier methods, such as sono-catalysis, homogeneous sonochemistry of liquids is also possible. In materials chemistry, mixed phase synthesis, and biological purposes, the investigation of chemical reaction improvement by ultrasonic has been conducted.

8.12. MODERN TRENDS

Chemical engineering will face new and concerning difficulties in the twenty-first century, including the need to use new methodological techniques, growing environmental concerns, and a shift in the raw material basis. But this century will also sneak into a lot of other industries, including biotechnology, pharmaceuticals, electronics, and other sectors, primarily food and agriculture.

The difficulties that sustainability of technological growth is confronting in a chemical sector revolve around reinventing the use of materials. To overcome these environmental issues, green chemical technologies are crucial. These technologies are therefore cost-efficient and environmentally friendly.

These ecologically friendly technologies are therefore affordable, practical, and suitable for the nation's geographical, social, ecological, technologically advanced, and economic characteristics. The only way to accomplish this goal may be through the development of new non-toxic, safe, and environmentally friendly products. In order to promote sustainable development and ensure better adherence to health, safety, and environmental regulations, the CPI must invest in green technologies.

ACTIVITY 8.1.

A food processing plant generates a large amount of waste that harms the environment. What sustainable biochemical technologies can be used to convert this waste into useful products, such as biofuels or biodegradable materials?

SUMMARY

The chapter "Sustainable Chemical and Biochemical Technologies" discusses various technologies that aim to reduce the negative impact of the chemical industry on the environment. The learning objectives of the chapter include understanding the development of chemical technologies, the importance of technological processes, the concept of biotechnology, the process of chemical nanotechnology, recent developments in green technology, and the use of alternative chemicals in the chemical industry

The chapter highlights the importance of developing sustainable chemical technologies that reduce waste, pollution, and energy consumption. It also discusses the use of biotechnology in producing chemicals and materials from renewable resources, such as biomass and waste. Furthermore, the chapter explores the potential of chemical nanotechnology, which involves the manipulation of matter on a nanoscale level, to develop new materials and processes that are more sustainable.

The chapter also discusses recent developments in green technology, including the use of renewable energy sources, such as solar and wind power, and the development of new materials that are biodegradable and recyclable. It also emphasizes the importance of using alternative chemicals, such as enzymes and natural compounds, in the chemical industry to reduce the use of hazardous chemicals.

Overall, the chapter highlights the importance of developing sustainable chemical and biochemical technologies to reduce the negative impact of the chemical industry on the environment and promote a more sustainable future.

REVIEW QUESTIONS

1. What are some of the latest chemical technologies in development and how do they compare to existing technologies in terms of sustainability and efficiency?
2. Discuss the importance of technological processes in achieving sustainable chemical and biochemical production. What are some of the key factors that determine the sustainability of a technological process?
3. Define biotechnology and explain how it is used in the production of sustainable chemicals and biochemicals. What are some of the advantages of using biotechnology over traditional chemical synthesis methods?
4. What is chemical nanotechnology and how is it used in sustainable production processes? Provide examples of its applications in the chemical and biochemical industries.
5. Describe some of the recent developments in green technology and their impact on the chemical and biochemical industries. How have these developments improved sustainability and reduced environmental impacts?

6. Discuss the use of alternative chemicals in the chemical industry and their potential benefits in terms of sustainability and environmental impact. What are some of the challenges associated with the use of alternative chemicals and how can they be overcome?

MULTIPLE CHOICE QUESTIONS

1. Which of the following is NOT an example of a chemical technology approach in development?

a. Carbon capture and storage
b. Biofuels
c. 3D printing
d. Electric vehicles

2. Why are technological processes important in chemical production?

a. They reduce production costs
b. They increase product yields
c. They improve product quality
d. All of the above

3. Biotechnology is the use of __________ to create products or improve processes.

a. Biological organisms
b. Chemical reactions
c. Physical properties
d. None of the above

4. Which of the following is a characteristic of chemical nanotechnology?

a. It involves the manipulation of materials on a nanoscale level
b. It is used primarily in the production of consumer goods
c. It does not require specialized equipment or facilities
d. All of the above

5. Which of the following is a recent development in green technology?

a. Carbon capture and storage
b. Hydraulic fracturing
c. Coal-fired power plants
d. Solar panels

6. **Alternative chemicals are used in the chemical industry because they:**
   a. Are cheaper than traditional chemicals
   b. Are safer for human health and the environment
   c. Have superior performance characteristics
   d. All of the above
7. **Which of the following is an example of a chemical technology approach in development for carbon capture and storage?**
   a. Using algae to absorb carbon dioxide from the atmosphere
   b. Sequestering carbon dioxide in underground geological formations
   c. Capturing carbon dioxide emissions and converting them to liquid fuels
   d. All of the above
8. **Biotechnology is used in the production of:**
   a. Pharmaceuticals
   b. Agricultural products
   c. Industrial enzymes
   d. All of the above
9. **Chemical nanotechnology has applications in:**
   a. Electronics
   b. Medicine
   c. Energy storage
   d. All of the above
10. **Which of the following is an example of an alternative chemical used in the chemical industry?**
   a. Petroleum-based solvents
   b. Chlorofluorocarbons
   c. Plant-derived surfactants
   d. None of the above

Answers to Multiple Choice Questions

1. (d)	2. (d)	3. (a)	4. (a)	5. (d)
6. (b)	7. (d)	8. (d)	9. (d)	10. (c)

REFERENCES

1. Butlin, J. (1989). Our common future. By World commission on environment and development.(London, Oxford University Press, 1987, pp. 383£ 5.95.).
2. Meadows, D. H., Meadows, D. L., Randers, J., & Behrens, W. W. (1972). The limits to growth. *New York*, *102*, 27.
3. Keeble, B. R. (1988). The Brundtland report:'Our common future'. *Medicine and War*, *4*(1), 17-25.
4. Houghton, J. T., Callander, B. A., & Varney, S. K. (Eds.). (1992). *Climate change 1992*. Cambridge University Press.
5. Grubb, M. (1993). The Earth Summit agreements: a guide and assessment; an analysis of the Rio'92 UN Conference on Environment and Development. Earthscan and the Energy and Environmental Programme of the Royal Institute of International Affairs.
6. Elkington, J. (1994). Towards the sustainable corporation: Win-win-win business strategies for sustainable development. *California Management Review*, *36*(2), 90-100.
7. Taniewski, M. (1999). Przemysłowa synteza organiczna. *Kierunki rozwoju, Wyd. Politechniki Śląskiej, Gliwice*.
8. Borowiecki, T., Gołębiowski, A., & Skowroński, B. (1998). Nowe procesy katalityczne w produkcji gazu syntezowego i wodoru z węglowodorów. Przemysł Chemiczny, 77, 128-132.
9. Taniewski, M. (2012). Chemistry of synthesis gas and carbon dioxide. The outline of contemporary opportunities. *Przemysl Chemiczny*, *91*(4), 492-498.
10. Taniewski, M. (2004). The challenges and recent advances in C1chemistry and technology. Polish Journal of Applied Chemistry, 48(1-2), 1-21.
11. Taniewski, M. (2006). Sustainable chemical technologies–development trends and tools. *Chemical Engineering & Technology*, *29*(12), 1397-1403.
12. Puchała, C. (2012). International action on chemical threats reduction. Chemistry Education in the Light of the Research. Pedagogical University: Krakow, 111-113.
13. Bednarek-Gejo, A., Mianowany, M., Skoczylas, P., & Głowacka, A. (2012). Świadomość ekologiczna studentów. *Hygeia Public Health*, *47*(2), 201-206.
14. Taniewski, M. (2005). Sustainable development in petrochemical industry. *Polish Journal of Applied Chemistry*, *49*(3), 183-194.
15. Taniewski, M. (2008). Sustainable chemical technologies in production of clean fuels from fossil fuels. *CLEAN–Soil, Air, Water*, *36*(4), 393-398.
16. Ismagilov, Z. R., Matus, E. V., Kerzhentsev, M. A., Tsikoza, L. T., Ismagilov, I. Z., Dosumov, K. D., & Mustafin, A. G. (2011). Methane conversion to valuable chemicals over nanostructured Mo/ZSM-5 catalysts. *Petroleum Chemistry*, *51*(3), 174-186.
17. Chum, H. L., & Overend, R. P. (2001). Biomass and renewable fuels. *Fuel Processing Technology*, *71*(1), 187-195.
18. Wilson, M. P., & Schwarzman, M. R. (2009). Toward a new US chemicals policy:

rebuilding the foundation to advance new science, green chemistry, and environmental health. *Environmental Health Perspectives*, *117*(8), 1202.

19. Bozell, Joe. (2001). Chemicals and Materials from Renewable Resources. 10.1021/bk-2001-0784.ch001. Anastas, P. T., & Kirchhoff, M. M. (2002). Origins, current status, and future challenges of green chemistry. *Accounts of chemical research*, *35*(9), 686-694.
20. Anastas PT, Zimmerman JB. Design through the 12 principles of green engineering. Environ Sci Technol. 2003 Mar 1;37(5):94A-101A. doi: 10.1021/es032373g. PMID: 12666905.Trost, B. M. (2002). On inventing reactions for atom economy. *Accounts of chemical research*, *35*(9), 695-705.
21. Metzger, J. O. (2006). Production of liquid hydrocarbons from biomass. *Angewandte Chemie International Edition*, *45*(5), 696-698.
22. Biermann, U., Bornscheuer, U., Meier, M. A., Metzger, J. O., & Schäfer, H. J. (2011). Oils and fats as renewable raw materials in chemistry. *Angewandte Chemie International Edition*, *50*(17), 3854-3871.
23. Guillena, G., Ramón, D. J., & Yus, M. (2007). Alcohols as Electrophiles in C C Bond-Forming Reactions: The Hydrogen Autotransfer Process. *Angewandte Chemie International Edition*, *46*(14), 2358-2364.
24. Chen, H., Armand, M., Demailly, G., Dolhem, F., Poizot, P., & Tarascon, J. M. (2008). From Biomass to a Renewable LiXC6O6 Organic Electrode for Sustainable Li-Ion Batteries. *ChemSusChem*, *1*(4), 348-355.
25. Lunsford, J. H. (1995). The catalytic oxidative coupling of methane. *Angewandte Chemie International Edition*, *34*(9), 970-980.
26. Lunsford, J. H. (2000). Catalytic conversion of methane to more useful chemicals and fuels: a challenge for the 21st century. *Catalysis Today*, *63*(2), 165-174.
27. Xu, Y., Bao, X., & Lin, L. (2003). Direct conversion of methane under nonoxidative conditions. *Journal of Catalysis*, *216*(1), 386-395.
28. Holmen, A. (2009). Direct conversion of methane to fuels and chemicals. *Catalysis Today*, *142*(1), 2-8.
29. Kioes, S., & Liebner, W. (2004). Methane-the promising career of a humble molecule. *Journal of Natural Gas Chemistry*, *13*(2), 71-78.
30. Koempel, H., Liebner, W., & Wagner, M. (2005). Lurgi's Gas to Chemicals (GTC): Advanced technologies for natural gas monetization. *Proceedings of the Gastech*.
31. Liebner, W., & Koempel, H. (2005, January). Gas to propylene: Report on commercialisation by Lurgi. In *18th World Petroleum Congress*. World Petroleum Congress.
32. Pushparaj, H., Ganesh, M. A. N. I., Muthiahpillai, P., Velayutham, M., Yong-Ki, P. A. R. K., Choi, W. C., & Jang, H. T. (2013). Effects of crystallinity of ZSM-5 zeolite on para-selective tert-butylation of ethylbenzene. *Chinese Journal of Catalysis*, *34*(2),

294-304.

33. Hongbo, R., Yuancheng, Q., Chengwei, S., Yutie, B., & Lin, Z. (2010). Adsorption and desorption properties of hybrid aerogels derived form MPMS-SSO at 77 K. *Rare Metal Mater Eng*, *39*, 475-478.
34. Suraweera, N. S., Albert, A. A., Peretich, M. E., Abbott, J., Humble, J. R., Barnes, C. E., & Keffer, D. J. (2014). Methane and carbon dioxide adsorption and diffusion in amorphous, metal-decorated nanoporous silica. *Molecular Simulation*, *40*(7-9), 618-633.
35. Rahmani, F., Haghighi, M., Estifaee, P., & Rahimpour, M. R. (2012). A comparative study of two different membranes applied for auto-thermal methanol synthesis process. *Journal of Natural Gas Science and Engineering*, *7*, 60-74.
36. Levenspiel, O. (2005). What will come after petroleum?. *Industrial & engineering chemistry research*, *44*(14), 5073-5078.
37. Jena, H. M. (2009). *Hydrodynamics of gas-liquid-solid fluidized and semi-fluidized beds* (Doctoral dissertation).
38. Fukui, T., Murakami, T., & Ichikawa, M. (1994). The behavior of ash components in the sludge melting process. *Water Science and Technology*, *30*(8), 197-207.
39. Nishikawa, G., Tyanagawa, H. M., Sakata, E., & Kojima, Y. (2001). Large Scale Urea Granulation Plants based on TEC Technology.
40. Marcilly, C. (2003). Present status and future trends in catalysis for refining and petrochemicals. *Journal of Catalysis*, *216*(1), 47-62.
41. Degnan, T. F. (2000). Applications of zeolites in petroleum refining. *Topics in Catalysis*, *13*(4), 349-356.
42. Degnan, T. F. (2003). The implications of the fundamentals of shape selectivity for the development of catalysts for the petroleum and petrochemical industries. *Journal of Catalysis*, *216*(1), 32-46.
43. Catani, R., Mandreoli, M., Rossini, S., & Vaccari, A. (2002). Mesoporous catalysts for the synthesis of clean diesel fuels by oligomerisation of olefins. *Catalysis Today*, *75*(1), 125-131.
44. Arpentinier, P., Basile, F., Del Gallo, P., Fornasari, G., Gary, D., Rosetti, V., & Vaccari, A. (2005). Role of the hydrotalcite-type precursor on the properties of CPO catalysts. *Catalysis Today*, *99*(1), 99-104.
45. Cavani, F., Centi, G., Perego, C., & Vaccari, A. (2005). Selectivity in catalytic oxidation: an issue or an opportunity for innovation?. *Catalysis Today*, *99*(1), 1-3.
46. Muda, A. A. B. (2014). Synthesis of Movtenb Oxide Catalysts for Propane Oxidation to Acrylic Acid via Microwave Irradiation Assisted Slurry Method.
47. Wang, H., Deng, Z., Chu, W., & Yang, W. (2009). A new approach to achieving a pure M1 phase catalyst for the selective oxidation of propane. *Reaction Kinetics and Catalysis Letters*, *97*(2), 233-241.
48. Van Hove, M. A. (2006). From surface science to nanotechnology. *Catalysis*

Today, *113*(3), 133-140.

49. Gawlitza, P., Braun, S., Leson, A., Lipfert, S., & Nestler, M. (2007). Herstellung von Präzisionsschichten mittels Ionenstrahlsputtern. *Vakuum in Forschung und Praxis*, *19*(2), 37-43.
50. Erdélyi, Z., Cserháti, C., Csik, A., Daróczi, L., Langer, G. A., Balogh, Z., & Erko, A. (2009). Nanoresolution interface studies in thin films by synchrotron x-ray diffraction and by using x-ray waveguide structure. *X-Ray Spectrometry*, *38*(4), 338-342.
51. Zhang, Z., Zhou, Y., Zhang, Y., Xiang, S., Zhou, S., & Sheng, X. (2014). Encapsulation of Au nanoparticles with well-crystallized anatase TiO 2 mesoporous hollow spheres for increased thermal stability. *RSC Advances*, *4*(14), 7313-7320.
52. Charpentier, J. C. (2005). Process intensification by miniaturization. *Chemical engineering & technology*, *28*(3), 255-258.
53. Sell, I., Ott, D., & Kralisch, D. (2014). Life cycle cost analysis as decision support tool in chemical process development. *ChemBioEng Reviews*, *1*(1), 50-56.
54. Francisco, É. C., Franco, T. T., Zepka, L. Q., & Jacob-Lopes, E. (2015). From waste-to-energy: the process integration and intensification for bulk oil and biodiesel production by microalgae. *Journal of Environmental Chemical Engineering*, *3*(1), 482-487.
55. Shanmugam, K., & Donaldson, A. A. (2015). Extraction of EPA/DHA from 18/12EE Fish Oil Using AgNO3 (aq): Composition, Yield, and Effects of Solvent Addition on Interfacial Tension and Flow Pattern in Mini-Fluidic Systems. *Industrial & Engineering Chemistry Research*, *54*(33), 8295-8301.
56. Noble, R. D., & Agrawal, R. (2005). Separations research needs for the 21st century. *Industrial & Engineering Chemistry Research*, *44*(9), 2887-2892.
57. Broering, J. M., Hill, E. M., Hallett, J. P., Liotta, C. L., Eckert, C. A., & Bommarius, A. S. (2006). Biocatalytic Reaction And Recycling by Using CO2-Induced Organic–Aqueous Tunable Solvents. *Angewandte Chemie*, *118*(28), 4786-4789.
58. Koros, W. J. (2004). Evolving beyond the thermal age of separation processes: membranes can lead the way. *AIChE Journal*, *50*(10), 2326-2334.
59. Davood Abadi Farahani, M. H., Rabiee, H., Vatanpour, V., & Borghei, S. M. (2016). Fouling reduction of emulsion polyvinylchloride ultrafiltration membranes blended by PEG: the effect of additive concentration and coagulation bath temperature. *Desalination and Water Treatment*, *57*(26), 11931-11944.
60. Dijkema, G. P. J., Ferrão, P., Herder, P. M., & Heitor, M. (2006). Trends and opportunities framing innovation for sustainability in the learning society. *Technological Forecasting and Social Change*, *73*(3), 215-227.
61. Lepoutre, J. (2008). *Proactive environmental strategies in small businesses: resources, institutions and dynamic capabilities*(Doctoral dissertation, Ghent University). White, A. (2009). 2nd Summit on the Future of Corporation: Paper Series on Restoring the Primacy of the Real Economy. *Retrieved August*, *24*, 2009.

62. White, A. (2009). 2nd Summit on the Future of Corporation: Paper Series on Restoring the Primacy of the Real Economy. *Retrieved August*, *24*, 2009.
63. Lorenz, P., & Eck, J. (2005). Outlook: Metagenomics and industrial applications. *Nature reviews. Microbiology*, *3*(6), 510.
64. Lorenz, P., & Zinke, H. (2005). White biotechnology: differences in US and EU approaches?. *Trends in Biotechnology*, *23*(12), 570-574.
65. Paula, L., & Birrer, F. (2006). Including public perspectives in industrial biotechnology and the biobased economy. *Journal of Agricultural and Environmental Ethics*, *19*(3), 253-267.
66. Taniewski, M. (2002). Alternative petrochemical technologies. *Przemysl Chemiczny*, *81*(11), 697-703.
67. Aramendía, M. A., Borau, V., Jiménez, C., Marinas, A., Marinas, J. M., Ruiz, J. R., & Urbano, F. J. (2004). Magnesium-containing mixed oxides as basic catalysts: base characterization by carbon dioxide TPD–MS and test reactions. *Journal of Molecular Catalysis A: Chemical*, *218*(1), 81-90.
68. Skutil, K., & Taniewski, M. (2006). Some technological aspects of methane aromatization (direct and via oxidative coupling). *Fuel Processing Technology*, *87*(6), 511-521.
69. Nouralishahi, A., Pahlavanzadeh, H., & Daryan, J. T. (2008). Determination of optimal temperature profile in an OCM plug flow reactor for the maximizing of ethylene production. *Fuel Processing Technology*, *89*(7), 667-677.
70. Sheldon, R. A. (1996). Selective catalytic synthesis of fine chemicals: opportunities and trends. Journal of Molecular Catalysis A: Chemical, 107(1-3), 75-83.
71. Sheldon, R. A. (1996). Chirotechnology: Designing economic chiral syntheses. *Journal of Chemical Technology and Biotechnology*, *67*(1), 1-14.
72. Sheldon, R. A., & Dakka, J. (1994). Heterogeneous catalytic oxidations in the manufacture of fine chemicals. *Catalysis today*, *19*(2), 215-245.
73. Sheldon, R. A., Wallau, M., Arends, I. W., & Schuchardt, U. (1998). Heterogeneous catalysts for liquid-phase oxidations: philosophers' stones or Trojan horses?. *Accounts of Chemical Research*, *31*(8), 485-493.
74. Sheldon, R. A. (2007). The E factor: fifteen years on. *Green Chemistry*, *9*(12), 1273-1283.
75. Sheldon, R. A. (2007). Enzyme immobilization: the quest for optimum performance. *Advanced Synthesis & Catalysis*, *349*(8-9), 1289-1307.
76. Sheldon, R. A. (2008). E factors, green chemistry and catalysis: an odyssey. *Chemical Communications*, (29), 3352-3365.
77. Sheldon, R. A., & Downing, R. S. (1999). Heterogeneous catalytic transformations for environmentally friendly production. *Applied Catalysis A: General*, *189*(2), 163-183.
78. Anastas, P. T., Heine, L. G., & Williamson, T. C. (2000). Green chemical syntheses

and processes: introduction.

79. Anastas, P. T., & Kirchhoff, M. M. (2002). Origins, current status, and future challenges of green chemistry. *Accounts of chemical research*, *35*(9), 686-694.

80. McDonough, W., Braungart, M., Anastas, P. T., & Zimmerman, J. B. (2003). Applying the principles of Green Engineering to cradle-to-cradle design. *Environmental Science & Technology*, *37*(23), 434A–441A. https://doi.org/10.1021/es0326322Anastas, P. T. (1998). Green Chemistry; Anastas, P. T., Williamson, TC, Eds. 0198502346, 9780198502340.

81. Sharma, S. K. (2010). Green Chemistry for Environmental Sustainability. Series: Advancing Sustainability Through Green Chemistry and Engineering. https://doi.org/10.1201/EBK1439824733.

82. Fortunak, Joseph. (2010). Eco-Friendly Synthesis of Fine Chemicals Eco-Friendly Synthesis of Fine Chemicals . Edited by Roberto Ballini (University of Camerino, Italy). From the RSC Green Chemistry Series. Edited by James H. Clark (University of York, UK) and George A. Kraus (Iowa State University, USA). Royal Society of Chemistry: Cambridge. 2009 . xii + 300 pp. $199.00. ISBN 978-1-84755-908-1 .. Journal of the American Chemical Society. 132. 10.1021/ja1000483.Soetaert, W., & Vandamme, E. J. (Eds.). (2010). Industrial biotechnology: sustainable growth and economic success. John Wiley & Sons. 10.1002/9783527630233.

83. Tang, W. L., & Zhao, H. (2009). Industrial biotechnology: tools and applications. *Biotechnology Journal*, *4*(12), 1725-1739.

84. Cheng, H., & Gross, R. (2011). Green Polymer Chemistry: Biocatalysis and Biomaterials (ACS Symposium). *American Chemical Society*. 10.1021/bk-2010-1043.ch001.

85. Hopwood, B., Mellor, M., & O'Brien, G. (2005). Sustainable development: mapping different approaches. *Sustainable development*, *13*(1), 38-52.

86. Lélé, S. M. (1991). Sustainable development: a critical review. *World Development*, 19(6), 607-621.

87. Sneddon, C., Howarth, R. *B., & Norgaard, R. B. (2006). Sustainable development in a post-Brundtland world. Ecological economics, 57(2), 253-268.*

88. Patil, K. D. (2013). Sustainable Green Chemical Technology for Developments in Process Industries. development, 27, 30.

89. *Patil, K. D. (2014). Review of Green Chemical Technologies for Sustainable Developments in Chemical Process Industries. Emerging Trends in Chemical Engineering, 1(2), 8-14.*

90. Irwin, A. (1995). *Citizen science: A study of people, expertise and sustainable development.* Psychology Press.

91. Renn, O. (2008). White paper on risk governance: Toward an integrative framework. In Global risk governance (pp. 3-73). Springer Netherlands.

92. Dalton, A. J. (1998). *Safety, health and environmental hazards at the workplace*. Cengage Learning EMEA.

93. Polshettiwar, V., Nadagouda, M. N., & Varma, R. S. (2009). Microwave-assisted chemistry: A rapid and sustainable route to synthesis of organics and nanomaterials. *Australian Journal of Chemistry*, *62*(1), 16-26.

94. Pace, V. (2012). 2-Methyltetrahydrofuran: A versatile eco-friendly alternative to THF in organometallic chemistry. *Australian Journal of Chemistry*, 65(3), 301-302.

95. Albrecht, M. A., Evans, C. W., & Raston, C. L. (2006). Green chemistry and the health implications of nanoparticles. *Green Chemistry*, *8*(5), 417-432.

96. Horváth, I. T., & Anastas, P. T. (2007). Innovations and green chemistry. *Chemical Reviews*, *107*(6), 2169–2173. https://doi.org/10.1021/cr078380vAnastas, P., & Eghbali, N. (2010). Green chemistry: principles and practice. *Chemical Society Reviews*, *39*(1), 301-312.

97. Li, C. J., & Trost, B. M. (2008). Green chemistry for chemical synthesis. *Proceedings of the National Academy of Sciences*, *105*(36), 13197-13202.

98. Anastas, P. T., & Kirchhoff, M. M. (2002). Origins, current status, and future challenges of green chemistry. *Accounts of Chemical Research*, *35*(9), 686-694.

INDEX

I

K

L

M

Q

R

U

V

W

X

Y

Kolibri 360 GmbH
Ferdinand-Jühlke-Straße 7
99095 Erfurt, Deutschland
produktsicherheit@kolibri360.de